PACKAGING LAW EUROPE

For M. and T., who have been with me since the beginning.

Packaging Law Europe

PATRICIA M. BAILEY
Member, State Bar of California
Doctoral Candidate, European University Institute, Florence
Of Counsel, Poitrinal & Associés / Galexia, Paris

Ashgate
Aldershot • Brookfield USA • Singapore • Sydney

Published by
Ashgate Publishing Ltd
Gower House
Croft Road
Aldershot
Hants GU11 3HR
England

Ashgate Publishing Company
Old Post Road
Brookfield
Vermont 05036
USA

British Library Cataloguing in Publication Data
Bailey, Patricia M.
Packaging Law Europe. - (Ashgate studies in environmental policy and practice)
1.Packaging waste - Law and legislation - European Union countries 2.Packaging waste - Environmental aspects - European Union countries
I.Title
341.7'62'094

Library of Congress Catalog Card Number: 99-72611

ISBN 1 85972 282 2

Printed and bound by Athenaeum Press, Ltd.,
Gateshead, Tyne & Wear.

Contents

Tables

Acknowledgements

This book arose from my LL.M. thesis at the University of Amsterdam and I would particularly like to thank Janet Griffiths, whom I met there and who has spent many hours rereading this text and providing me with both useful comments and encouragement. I would like to also thank my colleagues at the European University Institute for their support and encouragement, in particular, Anna Bernard, Andrew Butler, Mark Jeffery and Véronique Rideau. In Brussels, I was very fortunate to have access to materials and assistance from individuals working on packaging issues, among them Nancy Russotto, Charles Pirotte, Jacques Fonteyne, and the staff of ERRA and of Europen; I would like to thank them for their time and expertise. I am also very grateful to the editorial staff of Ashgate, Anne Keirby and Pauline Beavers, for their advice and patience with what turned out to be a much longer project than originally anticipated. Finally, I would like to thank Patricia Lebrat of Poitrinal & Associés, for the endless hours she has spent helping me prepare the manuscript.

Preface

Since the early 1970s, society has experienced a growing awareness of the problems caused by damage to the environment, followed by an escalating interest in protecting the environment for future generations. At the same time, packaging technology developed to meet the needs of the consumer society, such that by the late 1980s, the amount and visibility of packaging had increased exponentially. With a sense of both the need to protect the environment and as a back-lash against the excesses of the consumer age, the public began to call for regulation of both the amount of packaging produced and industry's methods of disposal. Much of this outrage was unfortunately based on emotion rather than science, as packaging waste actually makes up but a small fraction of the total waste produced. Be that as it may, the result of society's calls for action was a proliferation of packaging legislation that soon began to interfere with the free movement of goods in the Community. Thus, discussions began for the drafting of a directive regarding packaging at the European level. Due to the myriad of conflicting interests, this acrimonious debate continued for over four years; it ultimately resulted in the Directive on Packaging and Packaging Waste of December 1994. Given the contentious nature of the drafting of the Directive, it is hardly surprising that its implementation and enforcement have continued to be difficult.

This study addresses the enactment of the Directive and the ongoing disputes as a result thereof. It specifically looks at the debates surrounding constraints (such as reuse quotas and bans) on certain types of packaging, as well as the use of new instruments (such as eco-taxes and environmental agreements) to implement the Directive. The study demonstrates that the Directive was in fact, merely a 'Band-Aid' on the issue of how far Member States can go in protecting the environment versus the smooth functioning of the internal market, and concludes that given the continuing stand-off on this issue, it will be up to the European Court of Justice to determine the final parameters.

Introduction

In June of 1991, the German government enacted the *Verordnung über die Vermeidung von Verpackgungsabfällen,* (Ordinance on the Avoidance of Packaging Waste). The effects of this Ordinance set in motion a chain of events throughout the European Community that resulted in the enactment of new packaging legislation in some Member States and more stringent packaging legislation in others. The conflicts created by the divergent legislation led to calls for a directive at the Community level to harmonise the situation. After several years of discordant negotiations, on 20 December 1994, the European Parliament and Council finally enacted Directive 94/62/EC on Packaging and Packaging Waste ('the Directive'); nonetheless, the imprecision of the Directive which was necessary to ensure its enactment led to disputes which continue even today. At the heart of these disputes is the age-old struggle between the priorities of environmental protection and economic development.

Traditionally, packaging waste had been more of a reflection of consumer interests than environmental concerns. In the 1960s, consumers were mainly concerned with packaging which was easily disposable. In the 1970s, the energy crisis produced an emphasis on packaging which was as light as possible for ease of transport; industry accordingly focused on the use of lightweight packaging materials such as plastics. By the 1980s, the focus had shifted to issues such as extending product shelf-life via packaging materials, and tamper-proof packaging. By the 1990s, the attention of consumers had shifted once again: the issues of today are recycling and the recovery of packaging waste. The European Community produces about 50 million tons of waste each year from packaging materials alone; only about one-fifth of this waste is recycled. While Member States have historically dealt with packaging waste on an individual basis, the increasing mountain of waste resulting from packaging led to the enactment of the Directive.

This book analyses how the Directive has affected the European packaging landscape. In analysing the implementation of the Directive by Member States, consideration is given to such intra-European issues as the imposition of quotas on reusable containers; the use of economic instruments and environmental agreements; competition issues and state aids. The aim of the book is to enlighten the reader as to the current debates regarding packaging legislation which continue today, despite the enactment of Community legislation. The structure of the book is as follows:

Chapter 1 presents an overview of environmental law at the Community level with respect to packaging materials.

Chapter 2 examines the text of the Directive and looks at the compromises and amendments made during the drafting process. This chapter also addresses the work remaining to be completed by Community institutions.

Chapter 3 examines the types of packaging materials legislation implemented by the various Member States, in terms of the difficult systems used and the different allocations of responsibility along the packaging chain.

Chapter 4 looks at the ongoing debate regarding constraints on certain types of packaging, such as reuse quotas and bans, and considers the effects of such constraints on the internal market.

Chapter 5 analyses the use of economic policy instruments (eco-taxes) and environmental agreements to implement the Directive, as well as their effect on the internal market.

Chapter 6 considers competition issues raised by the Treaty and the Directive, as well as the use of commercial markings at the European level and the effects of state aids; it concludes with a look at general competition policy issues.

Appendix A is the complete text of the Directive, while Appendix B contains excerpts of other relevant legislation. Appendix C includes Commission Decisions on packaging issues that were issued subsequent to the Directive.

1 Environmental Legislation and Policy Regarding Packaging Waste at the European Community Level

An overview of environmental legislation and policy

Before considering the policy and legislation of the European Community on packaging waste, it is necessary to consider the evolution of European environmental policy and legislation as a whole. The Community's interest in environmental legislation can be traced back to 1972, the year of the United Nations Conference on the Human Environment; however, difficulties arose because the 1958 Treaty of Rome[1] (the 'Treaty') provides no clear basis for environmental protection. Instead, the Community began using Articles 100 and 235 (regarding the internal market) to legislate aspects of the environment which could arguably affect the free movement of goods or distort competition.

Such was the situation until 1987, when the Single European Act[2] ('SEA') amended the Treaty to add Articles 130r-130t. Article 130r formalised the adoption of both the 'precautionary principle' and the 'polluter-pays principle', which were already being applied in some Member States. Article 130r further states that environmental issues must become a standard part of other EC policies, including subsidiarity. Article 130s provides for the implementation of the policies outlined in Article 130r, in consultation with the European Parliament, but also provides for a unanimous vote in fiscal matters, matters of land use and measures significantly affecting the energy supply of a Member State. Article 130t notes that even when a measure is adopted pursuant to Article 130s, it shall not prevent a Member State from maintaining or introducing more stringent measures. Yet, despite the adoption of these articles, they were soon recognised as insufficient for the needs of the Community (Hooghe, 1993).

Accordingly, the Treaty on European Union[3] (the 'Maastricht Treaty') revised Articles 2 and 3 such that the environment is today given a status equal to that of economic concerns. Article 2 states that the Community is to pursue 'sustainable and non-inflationary growth, giving due consideration to the environment', while Article 3(k) empowers the Community to develop 'a policy in the sphere of the environment'. The addition of environmental issues to these first articles of the Treaty is more significant than the addition of Articles 130r-130t because these first articles address the basic functions of the Community. However, it is unclear whether the articles as revised will actually have a greater effect on the environment. Further, the Maastricht Treaty is based on the assumption that the single internal market has been fully achieved, a premise which is still open to debate.

In addition to Articles 130r-t, the SEA had added Article 100a, which permits the adoption of a measure by qualified majority of the Council, following deliberation with the European Parliament. Article 100a is an internal market provision, drafted to facilitate the passage of measures to ensure the smooth functioning of the internal market. Thus, by basing the Directive on Packaging and Packaging Waste[4] (the 'Directive') on Article 100a as opposed to Article 130s, the Community was declaring that the primary purpose of the Directive was to harmonise to the extent possible, the existing packaging legislation of the Member States (Reid, 1995).

During the discussions of proposed Community legislation, some Member States will view a proposed measure as being too harsh, while other Member States will view it as too weak. The end result is that the Commission will suggest a measure which is acceptable to the majority of Member States. In the field of the environment, the final legislation will rarely present the highest level of protection, but instead will reflect the current environmental and economic situation in the majority of Member States. It can be argued that the Commission must not propose a level of protection so lax that those Member States with existing legislation which imposes a higher level of protection will cause distortions in the internal market, post-enactment (de Sadeleer, 1995a). Indeed, this argument supports the German position regarding the Directive: prior to its enactment, Germany had existing legislation which provided a higher level of environmental protection and thus, Germany should be able to maintain that level, post-Directive. Yet, one of the primary reasons for the Directive (and why it was based on Article 100a and not Articles 130r-t) is that prior to the enactment of the Directive, national packaging treatment programmes, such as Germany's *Duales System Deutschland ('DSD')* were creating distortions in the internal market, such that an EC-wide directive became necessary.

Within Article 100a, paragraph 4 has become quite important, given that the opt-out provision of this paragraph indicates that Member States may apply national provisions in conflict with a directive, on the basis of the environment. However, the Member State must notify the Commission, which shall then confirm that the measure is not actually a form of arbitrary discrimination or a disguised restriction on trade. Thus, a Member State could actually have legislation more strict than the Directive. However, the amount of leeway granted is quite small and the restriction must meet the above criteria (de Sadeleer, 1995a).

The principle of subsidiarity

Another issue raised by the Maastricht Treaty is that of subsidiarity: the Community should take action 'only if and insofar as the objective of the proposed action cannot be sufficiently achieved by the Member States and can therefore, by reason of the scale of effects of proposed action, be better achieved by the European Community' (Article 3b). Thus, this principle restricts the power of the EC institutions to act in the environmental sphere when the Member States have sufficient competence. Yet, an alternative reading of this clause is that when an activity, due to the scale of its effects, cannot be sufficiently achieved by the Member States, it should be conducted by the Community. And in fact, the idea of 'reverse subsidiarity' was applied to the Community's packaging waste situation: the effects of national packaging waste programmes were sufficiently (and increasingly) European in scope such that it became necessary for the Community to step in, in order to secure the harmonisation of the internal market. More recently, the proposed 1997 Treaty of Amsterdam[5] would introduce a new restriction on the Community by which actions taken by the Community would not exceed that which is necessary to attain the objectives of the Treaty.

Article 171

The Maastricht Treaty contained a third element which can affect the environment: Article 171 was amended to include a provision for sanctions. Previously, the European Court of Justice ('ECJ') had no power to enforce its decisions, relying solely on the goodwill of the Member States. While this goodwill may have worked in some cases, certain Member States received several judgments against them for the same violation, but continually failed to comply. The ECJ can now impose penalties against a Member State for failing to implement its ruling. While this power will probably improve some Member States' compliance with the Directive, it will not make a difference for others. While the less-developed Member States have transposed most of the Directive and have established national waste programmes, several do not have the funding or technical training necessary to ensure compliance with many of the provisions of the Directive. Yet, to sanction these Member States or to reduce the funding they receive from the Community, will only set back their environmental progress further.

The Fifth Environmental Action Programme

The current Fifth Environmental Action Programme[6] diverges greatly from previous programmes in that in lieu of merely encouraging the protection of the environment, it approaches the problem as one of sustainable development. The Programme reverses its former perspective of a top-down approach and instead advocates the participation of those closer to the activities: the consumers, businesses and trade unions. As discussed in Chapter 3, the Netherlands and France were the first Member States to pro-actively include industry in the drafting of packaging legislation. In Europe, the approach to packaging materials legislation has been largely initiated by industry, or by government with industry as a partner to share the responsibility. This approach has emphasised reduction of the packaging before use, as opposed to other post-usage treatment such as recycling, reuse, incineration or landfill (OECD 1995), and is in line with the Community's policy of 'prevention' as the primary method of reducing packaging waste.

Community policy regarding packaging waste

The situation prior to the enactment of the Directive

The history of the Community's packaging waste policy is marked by three significant events.

The first event was precipitated by Denmark in 1981, when it passed legislation which banned all metal drink cans and required all beer and non-alcoholic drinks to be packaged under a deposit bottle system. In the resulting case, commonly known as 'Danish Bottles',[7] the European Court of Justice ('ECJ') determined that the restrictions placed by Denmark did not constitute a distortion of free trade and were appropriate and proportionate with regard to the environmental objectives. However, the ECJ did not approve of the Danish legislation's requirement that all drinks be sold in approved containers. While the *Danish Bottles* case has often since been cited as authority that Member States can place restrictions on the free movement of goods, in the name of an exemption for the environment, how far this exemption can be taken is another matter. Further, one must remember that the ECJ made its ruling in the absence of Community legislation on the matter; it is not clear that such restrictions would be upheld today.

The second significant event was the enactment of the forerunner to the Directive, the 1985 Directive on packaging for liquid foodstuffs[8] (the 'Liquid Foodstuffs Directive'). The need for this directive was due to the increase in packaging waste from liquids, as the result of the development of containers which made beverages more portable (and the proliferation of liquid container waste more visible). The Liquid Foodstuffs Directive focused on all containers (from manufacture to disposal) of liquid foodstuffs; Member States were required to set up programmes to reduce the weight and the volume of liquid container waste. The Liquid Foodstuffs Directive was not strongly supported by the Member States: by 1990, five of them had still not notified the Commission of their programmes. As the Liquid Foodstuffs Directive was implemented in a variable fashion from Member State to Member State, it ultimately began to create problems for the free movement of goods within the Community (Demey et al., 1996).

The third significant packaging event occurred in 1991, when the German government enacted the *Verordnung über die Vermeidung von Verpackgungsabfällen* (Ordinance on the Avoidance of Packaging Waste). This Ordinance affected all types of packaging materials and quickly set off a storm of controversy, both within and outside Germany. Debates soon began in other Member States regarding the enactment of their own packaging waste legislation. In some Member States (notably France and the Netherlands), legislation was enacted not long thereafter, and was the result of industry co-operating with government, in order to ward off strict legislation similar to that of Germany.

By 1993, problems created by the Member States' legislation had made it clear that intervention at the EC level was necessary, in order to remove barriers to trade and to prevent distortion of competition. However, strong differences of opinion existed as to what legislation was needed. Part of the problem was caused by the range of interests involved, such as packaging manufacturers, product manufacturers, retailers, consumers, and environmental interest groups (Long et al., 1997). In addition, there were differing national attitudes regarding the extent and gravity of the problem and how it should be addressed.

In drafting the packaging waste directive, the Commission encountered unprecedented conflicts, some of which were due to the disparate (and disputed) levels of waste generated by the Member States. As the following table makes clear, at the time of the enactment of the Directive, some Member States had more of a problem than others.

Table 1.1
Generation of municipal waste: kilograms per inhabitant per year (Non-EC countries have been included for comparative purposes)

EC Member States	
Finland	620
The Netherlands	500
Denmark	480
Luxembourg	450
Sweden	370
United Kingdom	350
Italy	350
Germany	350
Belgium	340
Austria	330
France	330
Spain	320
Greece	300
Ireland	310
Portugal	260
European Community (average)	350

Non-EC Countries	
United States	720
Hungary	460
Japan	410
Slovak Republic	360
Poland	340
Czech Republic	250

Source: *European Information Service* (1994a)

One must be careful however, not to make snap judgments from the above figures. While it appears that the wealthier countries overall tend to generate the most waste, Germany has one of the lowest rates, which one could attribute to their strong environmental awareness and the fact that *DSD* has been in effect for several years. On the other hand, Denmark and the Netherlands have rather high rates when one considers their strong recycling efforts. In fact, their rates are much higher than those of France or Spain, which are not known for their environmental awareness or recycling efforts. Even in Scandinavia, the rates are as low as 370 kg for Sweden and as high as 620 kg for neighbouring Finland. Moreover, a 1998 study by PriceWaterhouseCoopers revealed completely different statistics on waste generated by the same Member State, depending upon whether the source was the Member State's environmental ministry, the OECD, or Eurostat (Price, 1998). This finding underscores the need for the data collection provisions contained in the Directive.

Overall, one can conclude that the amount of waste generated depends not only on the Member State's relative wealth, but also on its production, distribution and consumption patterns, and recovery efforts. These disparate figures also suggest why the Directive had to be drafted in rather broad terms: it must allow flexibility for the varying needs of the Member States.

The enactment of the Directive

The enactment of environmental legislation

Although environmental legislation is of course, enacted in a fashion similar to that of other Community legislation, a few aspects are different. Proposals for legislation are made by the Commissioner in charge of the environment, following collaboration with the Commission, which is the only institution with the power to initiate new legislation. After the Commission makes a formal proposal, it must be approved by all of the Commissioners at their weekly meeting. The proposal is then transferred to the Committee of Permanent Representatives ('COREPER') for its review. Via this review (which is not open to the public) the Member States make actions to block or approve proposals (Hooghe, 1993).

Proposals regarding the environment are distinguishable in that they must be approved by the Environmental Council, which only meets twice per year. Thus, the legislative process is further retarded. Moreover, pursuant to Article 189b of the Treaty, proposals are subject to the co-decision procedure with the European Parliament. Most legislation provides for transposition into national law over a period of several years. Thus, a directive may take ten to fifteen years from the date of initial proposal to final transposition into national law. And only then do problems of enforcement begin.

The institutional process

Although an extensive discussion of the political history of the Directive is beyond the scope of this book,[9] an overview of the process is helpful in understanding the Directive's final form.

Although the original discussions regarding the Directive began in the late 1980s, the first draft of the Directive[10] did not emerge until late 1992. This first draft was highly criticised; at the time, the existing German packaging waste programme, *Duales System Deutschland,* was thought to be the best system and Member States were planning to each adopt a variation of it. However, as *DSD* began to have problems, particularly with the disposal of its plastic waste, critics began arguing that the Directive should take into account the various problems faced by *DSD*. In addition, it was clear from the beginning that it would be difficult to create successful packaging legislation at the European level, as the Liquid Foodstuffs Directive had been a failure due to poor enforcement (Dallemagne et al., 1993).

A particularly difficult issue to resolve was whether to have a target specifically for recycling, separate from that for recovery; a secondary issue was whether thereafter to have separate targets for plastics recycling, glass recycling, paper recycling, and so forth. The arguments were, not surprisingly, drawn between the north-south division in the Community, with the Netherlands, Germany and Denmark favouring high targets for recovery and recycling while Spain, Greece and Ireland protested vehemently. Ireland was particularly concerned because of its very limited sorting and recycling facilities, and made it clear that for the Directive to be successful, the Community would have to provide funding for such facilities (EIS, 1993).

During the 1993 debates, disputes also arose regarding the use of weight as opposed to volume, to set targets. By using weight, glass recycling would be favoured and a greater amount of light materials (such as plastics) would be landfilled. Unfortunately, despite advances in technology, plastics remained the most difficult material to recycle. Thus, a hierarchy of waste treatment also remained in contention, with Belgium and Germany arguing for a statement that prevention must be given highest priority (EIS, 1993).

In October of 1993, the proposed directive was amended[11] (the 'Amended Proposal') to reflect the input of government, industry and interest groups during the preceding year; yet, the targets remained largely unchanged. In December of 1993, the Environmental Ministers of the Community met and agreed on revised targets for the Directive. This accord was reached by majority vote, with Germany, Denmark and the Netherlands dissenting (Demey et al., 1996). By early 1994, despite the conflict the previous December, many parties involved with the Directive thought the issue was about to be resolved. However, with the accession of the environmentally-sensitive Member States of Austria, Finland and Sweden, the debate began anew, with these new Member States pressuring for the Directive to be rendered more environmentally sensitive by increasing the targets.

In March of 1994, the EC Ministers delivered their position on the Amended Proposal to the European Parliament. Groups such as the Alliance for Beverage Cartons and the Environment ('ACE') called on the European Parliament to approve the Amended Proposal, arguing that the revisions reflected the flexibility needed to accommodate the different markets in the Community. ACE and other industry interest groups also argued against discrimination between one-way versus refillable containers, noting that it remained unclear whether refillable containers were more environmentally friendly. The arguments of these groups were based in part on a report of the

Frauenhofer Institute in Germany, which had concluded that certain one-way packaging such as cartons could be recycled or incinerated with energy recovery at a lower cost to the environment than reusable plastic containers which, although refillable, must someday be landfilled. ACE and other groups also opposed the establishment of a hierarchy among options such as reuse, recovery, recycling and incineration (EIS, 1994b).

At the European Parliament, problems arose due to confusion regarding the co-decision procedure of the Maastricht Treaty. At the 4 May 1994 meeting of the European Parliament, 160 amendments to the draft were proposed, but not all were approved, in particular, those relating to the addition of a hierarchy, the imposition of greater producer responsibility and the proposed increases in the percentages of materials to be recycled. It bears noting that the 4 May 1994 vote was especially difficult, due to a prior contentious vote on another matter. By the time the packaging amendments came up for a vote, over 100 Members of European Parliament ('MEPs') had left the chamber, thus creating problems in obtaining the 260 minimum votes needed to pass each amendment. Thus, while some amendments received the requisite number of votes, others lacked the 260 votes needed, yet received approval from an overwhelming majority of MEPs present (EIS, 1994c).

On 25 May 1994, when the European Commission received the votes from the Parliament, it took the unusual step of adopting all of the proposed amendments, including those that had not received sufficient votes; the proposed hierarchy among the methods of waste treatment was however, only added to the Preamble. But this step angered Belgium, which was extremely concerned about a proposed amendment to Article 15, regarding economic instruments. Despite the opinion of the Commission's legal service that the proposed amendment would not allow a challenge to Belgium's fledgling eco-tax programme, Belgium refused to approve any of the amendments (EIS, 1994d).

The Commission then presented the proposal to the Environment Council on 13 July 1994. It was at this point that Germany, Denmark and the Netherlands indicated that they would vote against the proposal, although only three negative votes would not prevent the Environment Council from obtaining the qualified majority needed to approve it. Thus, in the 6 July 1994 meeting of the COREPER, agreement could not be reached on a final version of the proposed directive to present to the European Council. Accordingly, the entire matter had to be determined by the Conciliation Procedure (EIS, 1994e, 1994f).

Yet, even the Conciliation Procedure became embroiled in disputes as ministers and MEPs fought over the wording of Article 15 or even the intricacies of the rules of order (the Directive was the first piece of legislation to which the procedure had been applied). Of particular concern to the MEPs was the amendment to Article 15 that economic instruments 'shall neither harm competition nor create any obstacles to freedom of movement, nor cause discrimination against imported goods'. Although this phrase appears to be a reiteration of one of the fundamental principles of the Treaty, some Member States feared it would endanger the eco-tax programmes (Demey et al., 1996; EIS, 1994g, 1994h). The conciliation committee worked intensively during the autumn of 1994 in order to reach an agreement regarding Article 15, by which the above provision was deleted and instead, Member States were permitted to adopt economic instruments 'in accordance with the principles governing Community environmental policy, *inter alia,* the polluter-pays principle, and the obligations arising out of the Treaty...'. The definitive version of the Directive, which contained the latter provision, was submitted to the Council of Ministers on 14 December 1994, at which time it was approved by a qualified majority, with Denmark, Germany and the Netherlands dissenting (EIS, 1994h, 1994i). The final provisions of the Directive thus reflect the numerous compromises necessary to accommodate not only environmental concerns, but also competition and trade law issues.

How the Directive is a reflection of Community environmental policy

The final version of the Directive does reflect the goals of EC environmental policy in several respects. For example, paragraph 3 of Article 130r requires that the Community take into account four criteria during the enactment of all legislation. The first of these criteria is that of 'available scientific data'; accordingly, Article 19 of the Directive provides for the consultation of a scientific committee and for the amendment of the Directive as necessary, in order to reflect scientific and technological progress. As the demand for waste treatment grows as the result of the Directive, technology will changed significantly and thus, this article will become of vital importance. The second requirement is that the Community take into account the environmental conditions in the various regions of the Community; in this respect, the Directive provides for delayed implementation by Portugal, Greece and Ireland, which were thought to have more difficulties in this area, due to the insufficiencies of then-existing waste treatment plans.

The third criteria of Article 130r is that the Community consider the potential benefits and costs of action or lack of action. With regard to the Directive, it was clear that the cost of inaction or failure to harmonise the disparate packaging waste laws could be quite severe, particularly when one considers the example of the scrap paper market in the United Kingdom collapsing, due to the influx of paper from Germany as the result of *DSD.* Thus, the benefit to be gained from the Directive is not only a reduction of the packaging waste produced in the EC, but also a reduction of the effects of packaging waste treatment on the internal market. The final criteria of Article 130r is the consideration of the economic and social development of the Community and the balanced development of the regions. The Directive seeks to ensure that citizens in the Mediterranean countries (which previously had little or no packaging legislation) are granted the same level of packaging waste treatment as the countries of Northern Europe. In addition, Article 6(6) of the Directive seeks to assure that Member States with extensive packaging waste treatment will only export their waste to Member States which have sufficient capacity to handle the exports.

Yet, while the foregoing demonstrates the Community's commitment to the environment, it must be remembered that the Directive is based on Article 100a. Thus, while the effect of the Directive is the protection of the environment, its purpose is to avoid distortions of the internal market. Indeed, it is with respect to the internal market that many of the disputes have arisen.

Notes

1 Treaty establishing the European Community, signed at Rome on 25 March 1957.
2 Single European Act, signed at Luxembourg on 17 February 1986.
3 Treaty on European Union, signed at Maastricht on 7 February 1992.
4 European Parliament and Council Directive 94/62/EC of 20 December 1994 on packaging and packaging waste, *Official Journal* No. L 365, 31†December 1994.
5 Treaty of Amsterdam, signed at Amsterdam on 2 October 1997, but not yet fully ratified by all Member States.
6 Resolution of the Council of the European Communities and of the Representatives of the Governments of Member States meeting within the Council, of 17 May 1993, on the continuation and implementation of a European Community policy and action programme on the environment (1993 to 1998), *Official Journal* C138/1 (1993).
7 The Commission of the European Communities *v* The Kingdom of Denmark [1988] ECR 4607. This case is further discussed in Chapter 4.
8 Council Directive 85/339/EEC of 27 June 1985 on packaging for liquid foodstuffs, *Official Journal* No. L176, 6 July 1985.
9 See Eden, 1996 and Haverland, 1998.
10 Proposal for a Council Directive on packaging and packaging waste, *Official Journal* No. C 263, 12 October 1992.
11 Amended proposal for a Council Directive on packaging and packaging waste, *Official Journal* No. C 285, 21 October 1993.

2 The Packaging Waste Directive

Introduction

The enactment of the Packaging Waste Directive[1] ('the Directive') took more than five years to complete. Like most legislation, the final version of the Directive is a reflection of the influence of different governmental institutions, as well as NGOs (including trade associations and environmental interest groups) and the general public. However, the Directive was one of the most contentious pieces of legislation ever enacted by the Community. The controversy was partly caused by the growing influence of the European Parliament, with several members becoming extensively involved in the process. Another reason was simply that packaging is something to which everyone can relate: it affects the lives of citizens in all social classes, at all income levels, and on a daily basis. In fact, packaging affects more than half of the goods transported within the Community and has a greater effect on municipal waste than all other groups of manufactured goods (Long et al., 1997). Thus, the Directive has had not only environmental, but financial and social repercussions as well.

Several years after its enactment, reaction to the Directive has been mixed. While industry has generally been positive because uniform packaging regulations make it easier to market a product throughout the Community, the inconsistent implementation of the Directive by different Member States continues to create problems. From another perspective, environmentalists have been critical because the Directive did not maintain the strict requirements which were already in effect in several of the Member States, notably Germany and the Netherlands.

One can divide the provisions of the Directive into four groups: those placing an obligation on the Member States; those placing an obligation on the Community institutions; those which grant Member States latitude to act in certain areas; and those requiring follow-up activities (Demey et al., 1996).

Yet, it is useful to first look at the scope of the Directive and its definitions, which themselves have been criticised as being too vague.

General principles

Scope and objectives

The scope of the Directive is quite broad; this reflects the Preamble's emphasis that a sectoral approach to the treatment of packaging materials had not been successful and that a new approach had to be undertaken at the EC level. The Directive applies to all packaging materials in the Community, including imported packaging and packaging materials, at both the retail and wholesale levels. The Directive also applies to all commercial, industrial and transport packaging (excluding shipping containers). The Directive even applies to packaging materials for the medical and pharmaceutical markets (Articles 2 and 3). The final version of the Directive emphasises that it shall not adversely affect either existing Community legislation regarding safety, health and hygiene or legislation regarding the transport of hazardous wastes. Regardless of the sector involved, the Directive applies to packaging materials at the primary, secondary and tertiary levels, as defined *infra*.

Article 1 states that the Directive's objectives are to harmonise Member States' packaging laws in an attempt to reduce the impact of packaging materials on the environment, and to secure the functioning of the internal market by preventing obstacles to free trade and distortion of competition within the Community. The importance of reducing interference with the internal market is emphasised by the fact that the Directive is based on Article 100a of the Treaty (harmonisation) as opposed to Article 130s (environmental protection). The initial proposal for the Directive[2] ('the Initial Proposal') stated that it would meet its objectives by establishing: 1) essential requirements and targets for packaging materials; 2) measures to prevent packaging waste; and 3) methods for proper waste management. The final version of the Directive instead identifies a first priority of reducing the production of packaging waste overall, with additional 'fundamental principles' of reusing and recovering packaging through recycling and other forms of recovery, the ultimate goal being to reduce the amount of packaging which is treated via landfill or other methods (Article 1). Germany, Luxembourg and the Netherlands had favoured a more strict version of this article which would have established a hierarchy of prevention and would have more clearly defined the principle of packaging responsibility. However, other Member States refused to allow a hierarchy to be established, having determined that the existing technology remained too undeveloped to conclude that one method should be favoured (EIS, 1993). This refusal was indeed quite prescient, given that in recent years, certain Member States have tried to claim that 'reuse' is a form of 'prevention'.

The Preamble states that deviations from the target ranges can be permitted to the extent they do not interfere with the workings of the internal market. Accordingly, the Preamble acknowledges that certain Member States will be permitted to maintain targets which reflect a high level of environmental protection so long as these targets do not create barriers to trade, whereas other Member States will be permitted to have lower targets with the proviso that they must achieve both minimum targets for recovery within a specified deadline and the Directive's targets by a later deadline. The Preamble also emphasises that efforts in one Member State to improve the environment must not impede the efforts of another Member State to comply with the Directive.

Definitions

Prior to the enactment of the Directive, a recurring problem in conducting comparisons among different packaging waste legislation was the disparate range of terminology among (and even within) the Member States. Such varied definitions can lead to significant misinterpretations of the statistics (OECD, 1996). The establishment of common definitions should thus improve the implementation and enforcement of the Directive, despite the fact that some of them are rather vague.

Packaging classification Article 3.1 of the Directive begins with a broad definition of 'packaging':

> all products made of any materials of any nature to be used for the containment, protection, handling, delivery and presentation of goods, from raw materials to processed goods, from the producer to the user or the consumer. 'Non-returnable' items used for the same purposes shall also be considered to constitute packaging.

The Directive thereafter identifies three groups of packaging: primary, secondary and tertiary. 'Primary packaging' encompasses all packaging which is transferred to the consumer as part of the sales unit. This definition includes packaging used for marketing purposes, such as cartons enclosing tubes of toothpaste (Article 3.1(a)). In certain Member States, such as Germany, strict packaging regulations have resulted in the near disappearance of this type of packaging as manufacturers have sought to reduce the primary packaging to only that which is essential for the use of a product, or for the safety of the consumer. For example, containers of aspirin may still be sold in boxes because the carton represents an additional barrier against tampering, whereas external packaging for personal care items (such as facial creams, deodorants and shampoos) has been virtually eliminated. The need to reduce primary packaging at the source has also led to many innovations in the detergent industry, where it is now possible to buy refillable bottles and concentrated powders.

The Directive uses the term 'secondary packaging' to refer to packaging used for grouping products together. This definition applies to packaging used for ease of shipment or to facilitate in-store stocking (such as carton dividers). In addition, secondary packaging includes packaging which is used to hold together a set of items at point of sale and which may or may not be delivered to the consumer. The Directive further defines secondary packaging as that which 'can be removed from the product without affecting its characteristics' (Article 3.1(b)).

Lastly, 'tertiary packaging' is defined as packaging used to transport the product. This definition includes packaging used to facilitate transport as well as that used to prevent mishandling or damage. However, the final version of the Directive specifically excludes containers used for air, sea and land transport (Article 3.1(c)).

Interestingly, the amendments to the proposed Directive[3] ('the Amended Proposal') contained a category for 'small packaging' which included 'all primary packaging with a weight of three grams or less and a volume of 100 cm^3 or less'. This provision was deleted from the final version; Article 20, which was substituted in its place, provides for 'Special Measures' for small packaging (which are not defined). Special Measures may also be provided for luxury packaging and for packaging in the pharmaceutical and medical device industries. The latter has unique considerations because it must take into account the existing legislation in the Member States regarding safety, performance and quality, as well as other Community legislation regarding medical devices. Special Measures are to be approved via the Committee Procedure, *infra*.

Packaging waste Article 3.2 defines 'packaging waste' as any packaging or packaging materials (excluding production residues) contained in the definition of 'waste' in the Waste Directive;[4] this directive defines 'waste' as any substance or object in selected categories, which the holder discards or intends or is required to discard. A review of these categories makes it clear that almost any type of packaging or packaging material will fall under the category of 'waste' and thus be subject to the Packaging Waste Directive. Article 3.10 likewise defines 'disposal' to include any of the many operations listed in the Waste Directive. In a similar fashion, the definition of ‘packaging waste management’ is taken from the Waste Directive, namely, ‘the collection, transport, recovery and disposal or waste, including the supervision of such operations and the after-care of disposal sites’ (Article 3.3).

Article 3.4 defines ‘prevention’ as the ‘reduction of the quantity and of the harmfulness for the environment of [both]: materials and substances contained in packaging and packaging waste, [and] packaging and packaging waste at production process level and at the marketing distribution, utilisation and elimination stages’. The Directive concludes that this reduction shall be achieved via the development of ‘clean products and technology’.

Recycling versus Recovery Recycling and recovery are significant terms because they affect the ability of Member States to meet the targets of the Directive. ‘Recycling’ is the ‘reprocessing in a production process of the waste materials for the original purpose or for other purposes including organic recycling *but excluding energy recovery’* (Article 3.7, emphasis added). What this definition means is that packaging waste may be affected via a production process (such as sterilisation) so that it can be used again (as either a package or for other purposes) and may even be used again in an organic recycling process, as discussed *infra.* However, recycling specifically excludes any reprocessing of packaging waste for energy recovery, such as incineration.

Article 3.6 states that ‘recovery’ encompasses any of the operations listed in Annex II B of the waste Directive.[5] Thus, ‘recovery’ includes incineration (‘use primarily as a fuel or other means to generate energy’) while ‘recycling’ does not. The exclusion of incineration as a form of recycling has become significant in light of the separate targets set for recycling versus recovery. In the early 1990s, an intense debate arose in Germany regarding the exclusion of incineration as an acceptable form of recovery. In other Member States, as the packaging waste legislation debates began, it soon became clear that many Member States would be unable to meet targets for recovery without the use of incineration (Emballages, 1992a). In December of 1993, following extensive discussions, the EC Environmental Ministers voted to permit a limited amount of incineration as a form of recovery in the final version of the Directive (Cairncross, 1993). Today, even Germany permits incineration, subject to certain provisions: energy recovery must be the primary focus of the incineration; the energy value must be at least 11,000 KJ/kg for each material; the combustion efficiency must be 75% or more; and the energy recovered must actually be used (ERRA, 1998).

The final version of the Directive clarified the definition of ‘organic recycling’ so that it now comprises the ‘aerobic (composting) or anaerobic (biomethanization) treatment, under controlled conditions and using micro-organisms, of the biodegradable parts of packaging waste, which produces stabilised organic residues or methane' (Article 3.9). This article emphasises however, that landfilling cannot be considered as organic recycling.

Implementation Date Article 22 of the Directive requires Member States to enact all legislation necessary for compliance on or before 30 June 1996 (the 'Implementation Date'). The Implementation Date is the pivotal date from which many deadlines of the Directive havc been calculated.

Obligations upon the Member States: quantitative aspects

One of the most perplexing problems during the drafting of the Directive was how to harmonise the existing packaging laws of the few Member States which had them. Eventually, it was concluded that due to the differing consumption rates and differing types of packaging consumed in the Member States, it would be impossible to identify one system or method which would work throughout the Community. In the end, as so often happens during the drafting of legislation, compromise was necessary in order to enact any type of law. Yet, while the final version of the Directive is more imprecise than originally envisaged, certain harmonisation measures were eventually made.

Targets for recovery and recycling

A major focus of the Preamble is to promote the free flow of goods throughout the Community and to prevent barriers to trade. The Preamble notes that the then-existing differences in the packaging policies of Member States (which included disparate targets for recovery and recycling, varying requirements for the essential composition of packaging materials, and dissimilar policies regarding acceptable forms of recycling and recovery) were interfering with the internal market. Hence, similar target ranges for recycling and recovery needed to be established for the entire Community. (A previous draft of the Directive had called for harmonisation of all packaging targets and measures, but agreement as to which level could not be achieved.) One of the most controversial aspects of the Directive was thus the designation of targets for the recovery and recycling. As one might expect, these targets underwent the radical transformation during the negotiations.

Original targets for recovery and recycling The Initial Proposal identified final targets to be met within ten years: only 10% of the packaging materials would be dispensed with via a final disposal method (such as landfilling), while the remaining 90% of the packaging material would be recovered. Of that 90%, the Initial Proposal required that 60% of the weight of each material in the package was to be recycled. The Amended Proposal of 1993 provided that within five years of implementation, intermediate targets of 60% of the packaging materials for recovery and 40% of the recovered packaging materials for recycling were to be met.

Final targets for recovery and recycling The targets in the final version of the Directive were reduced substantially. The Directive designates targets which must be met within five years of the Implementation Date, i.e., 30 June 2001: first, a minimum of 50% and a maximum of 65% of the weight of packaging waste must be recovered. Second, of the materials recovered, a minimum of 25% and a maximum of 45% by weight of the total of packaging materials contained in a Member State's packaging waste must be recycled. Of the materials recycled, at least 15% by weight of each packaging material must be recycled (Article 6.1). In a controversial vote during the negotiation of the Directive, the European Parliament voted against raising the recycling requirement to 25%, allegedly due to pressure from the influence of the plastics industry (EIS, 1994c). However, this position is validated by the fact that the use of weight and not volume leads to a favouritism toward other materials (such as glass) which are not necessarily more environmentally-friendly. Moreover, the provision for each material is significant because otherwise, the manufacturer could achieve the recycling target by recycling more of other materials which comprise the package. The end result would then be that technologies for the disposal of certain materials would never be developed as long as industry could comply with the Directive by recycling other materials.

The targets in the final version of the Directive reflect extensive compromise by the northern Member States to obtain at least some type of packaging legislation. Further, the experiences of packaging manufacturers during the early years of some Member States' programmes made it clear that European-wide legislation was needed; thus, the political pressure for compromise was strong. Even the Amended Proposal still reflected the strong, early sentiments of Member States, when the German minister, Klaus Töpfer, wanted to require an 80% recycling rate of all packaging materials by the end of 1995. Yet, by late 1993, the northern Member States had been forced to make such concessions that *Emballages,* the packaging trade journal of France, noted, *'Les pays du Nord ont du mettre beaucoup d'eau dans leur vin.*[6] In the end, even further compromise was needed in order to obtain a consensus on the Directive. However, total consensus was never reached as Germany, Denmark and the Netherlands ultimately voted against the final version of the Directive, arguing that it was too weak to be effective (EIS, 1994j). What made it so difficult to find a solution was the diversity of opinions regarding targets, which were a reflection of the disparate levels of environmental awareness and the varying recycling and recovery capabilities among the Member States.

Review of the targets The Directive provides that every five years, the Council shall review the targets for recycling and recovery. The initial review is to be concluded using a Commission report, which is to be based on the results of scientific research and which must be completed by 30 June 2000. No later than 1 January 2001, the Council must set the targets for the second five-year period; the Council is required to act by qualified majority and upon a proposal from the Commission. Thereafter, the reviews shall take place every five years. The purpose of the reviews is to eventually raise the targets substantially (Article 6). Moreover, the review power permits the Commission and the Council to exert pressure on any sector or materials for which the targets are not being met.

Criticisms of the targets While the Directive has been criticised for failing to provide sufficient guidelines to the Member States, the intent of the Directive was clearly to leave the methods of achieving the goals to the Member States, as opposed to prescribing a specific system. Yet, the targets of the Directive are themselves questionable because they reflect traditional assumptions which favour the use of glass and metal over plastic; some studies suggest that with the exception of PVC, lighter-weight plastic packaging may be the most ecologically efficient (Europen, 1996). Moreover, the volume of the packaging tends to be more significant than the weight when one looks at the effects of packaging on the infrastructure of waste collection and landfill capacity (OECD, 1996). Fortunately, Article 19 makes provision for scientific advancements by permitting not only the targets, but also the identification system and the database to be modified ad hoc, in order to reflect technological developments. Such amendments are to be made via the Committee Procedure, *infra*.

Provisions for Member States unable to meet the targets Article 6.5 makes special provisions for Greece, Ireland and Portugal. These Member States are not required to attain the final targets for recovery and recycling until five years after the Implementation Date, i.e., 30 June 2001. In the interim however, these Member States must achieve a recovery rate of at least 25%; they may also delay the date by which the final targets of the Directive are achieved, but not beyond 31 December 2005.

The rationale for the foregoing exemptions was that the specific situations of these Member States, namely, the presence of rural and mountainous areas in Ireland, the extensive number of small islands in Greece, and the low level of packaging consumption in Portugal made it more difficult for these Member States to implement the Directive (Article 6.5). However, the underlying reason is more likely due to the political efforts of these Member States. During the drafting process, it was argued that these Member States had such limited sorting and recycling facilities that it would not be possible for them to meet even the lowest recycling targets within the deadline. The extension of the deadline and the reduction of targets have given these Member States the time needed to develop their waste treatment industries.

Provisions for Member States with more stringent targets The Directive provides that in the interest of attaining a high level of environmental protection, Member States are permitted to have targets which exceed those of the Directive, subject to several provisos. The first proviso is that the Member State have 'appropriate capacities' for recycling and recovery (Article 6.6). The use of the phrase 'appropriate capacities', as opposed to 'sufficient capacities' is significant because it means that if a Member State has insufficient capacity, it may export packaging waste to another Member State for treatment and disposal.

Which thus leads to the second proviso of Article 6.6, that the Member States are permitted to have programmes which exceed the targets of the Directive, to the extent the programme does not result in distortions to the internal market or hinder compliance by other Member States. Thus far, Austria, Belgium and the Netherlands have notified the Commission of targets in their legislation which surpass those of the Directive. The Commission has asked other Member States for their views via the Committee Procedure and a decision on the requests of these three Member States is expected by early 1999. A Member State which objects to an Article 6.6 notification must provide substantial proof that the excess waste will adversely impact another Member State *at the time of the application.* A future adverse impact would not be a ground for denial, although the Commission could insert in its approval a proviso that the permit could be rescinded in the future, should the higher targets have such an impact.

Indeed, while it is clear that a Member State must have appropriate capacity, this capacity may be met anywhere in the Community or outside it as well. As the European Court of Justice has ruled that waste is a tradable good, Member States have the right to ship their packaging waste anywhere outside their territory for treatment. And in fact, some Member States already ship large quantities of waste for treatment outside the Community, to either Eastern Europe or countries of the third world (de Sadeleer, 1995a). From a policy standpoint, such a proposition, although perfectly legal, would be in contradiction with the principle of proximity, by which wastes should be disposed of as close to source as possible. Recently, France received a

Reasoned Opinion from the Commission for failure to comply with Regulation 259/93 concerning the monitoring of waste shipments in the Community. France refused to accept plastic waste from Germany, which waste it considers to be on the Regulation's Amber List (household waste) and which may be rejected by the importing country, while Germany and the Commission consider such waste to be on the Green List, which signifies that it can circulate freely (EIS, 1998a).

Return, collection and recovery systems

Article 7 is the crux of the Directive, stating what Member States must do to packaging waste in order to comply with the Directive. Member States must ensure that system(s) are established to provide for the return and collection of both used packaging and packaging waste from the final user, whether it be the consumer or some other individual/entity. The system established must then ensure that the waste is directed to 'the best waste management alternatives'. Which of the current alternatives will be utilised is left to the particular Member State to decide. The Amended Proposal added the requirement that the packaging materials must be recovered from all points in the waste stream; this places an additional burden on the economic operators to obtain the used packaging materials from all sources, including the retail stores (as has been the case in Germany since 1992).

In order to meet the targets of the Directive, Member States must also establish systems to ensure the proper reuse and recovery of the packaging materials (Article 7.1). The methodology of these systems is left to the discretion of each Member State, but it must ensure that the system is open not only to the economic operators within its territory, but also to operators from other Member States. The appropriate public authorities must also be involved.

Member States are allowed to apply the requirements and tariffs of the recovery systems to packaging imported from other Member States and from countries outside the EC. However, fees levied on packaging from other Member States must not discriminate in favour of packaging from the Member State implementing the system and the tariffs must not distort competition within the Community (Article 7.1). This article implies that in the absence of co-ordinated agreements among the packaging waste authorities (such as that between *Eco-Emballages* and *Duales System Deutschland* to mutually respect their 'Green Dots'), a manufacturer must ensure the recovery of its packaging waste by executing an agreement with the packaging waste recovery authorities of each Member State where it intends to sell its products. In addition, a logical consequence of Article 7.1 is that Member States are free to charge higher tariffs on packaging from outside the Community; yet, it is likely in such a case that non-EC countries would retaliate by imposing higher tariffs on packaging from the Community.

The Directive also provides that the return, collection and recovery systems are to be established as part of a policy regarding packaging materials; such policy must consider not only consumer protection, health, safety and hygiene, but also the technical characteristics of the materials used. Nonetheless, all industrial property rights must be respected (Article 7.2).

The Amended Proposal had contained a broad requirement that Member States must take special account of the problems the Directive might cause for small to medium-sized enterprises (SMEs). Unfortunately, this requirement was deleted from the final version of the Directive and instead, Member States are merely obligated to assist SMEs with the provision of information regarding the amounts of packaging produced and recovered. Unfortunately, SMEs already have the greatest difficulty complying with the packaging laws of Member States. For example, if a small Italian company wishes to export its products to France, it must become a member of *Eco-Emballages* or risk its products being refused by French retailers. While the expense of compliance may result in only a small increase in the cost of the product, the cumulative effect of requiring an SME to join the programme of several Member States is quite severe.

Obligations upon Member States: essential requirements

Requirements of Article 9 regarding the composition of packaging

The Preamble notes that toxic and/or noxious materials in packaging often leach into soil when landfilled or escape into the atmosphere when incinerated; it proposes to eliminate these materials as much as possible before they become part of the packaging, by having a technical committee set standards regarding the acceptable composition of a package. The Directive thus sets forth essential requirements with which the Member States must comply. First, by 31 December 1997, Member States were required to have taken measures to ban any packaging which did not comply with the essential requirements. Member States may however, presume compliance when the packaging materials are in conformity with either harmonised standards set forth in the *Official Journal,* or with the applicable national standards if no harmonised standards exist. Each Member State must notify the Commission of national standards which it believes are in compliance with the requirements of the Directive. These standards will then be transmitted to other Member States and published in the *Official Journal*. Should a dispute arise as to whether a Member State's standards are in compliance, the issue is to be determined pursuant to Directive 83/189/EEC.[7] The Commission will then inform the Member State as to whether it is necessary to withdraw the standard. The Member State can thereafter challenge the Commission's decision before the European Court of Justice.

Requirements of Annex II

In three sections, Annex II provides the specifications as to the composition, the reusability and the recyclability of packaging materials.

Composition and manufacture of packaging materials Packaging is to be manufactured such that safety, hygiene and marketability are attained with the minimum volume and weight of the packaging materials. This requirement is in accordance with the principal of prevention. Packaging must be designed for optimal reuse and recovery (including recycling), such that the portion of materials which must be disposed of via landfill is minimal. Further, packaging is to be designed to contain the minimum amounts of hazardous materials possible, with respect to the presence of hazardous materials in emissions, ash or leachate during the incineration or landfill processes (Annex II, para. 1).

Specifications as to the reusability of packaging materials With respect to the reusable nature of packaging, three requirements must be simultaneously satisfied. First, it must be possible to process the packaging materials so that they meet health and safety requirements (for example, liquid drink containers which can be sterilised). Second, the packaging materials must enable the package to be reused after being subject to customary conditions of use. Third, the packaging materials must fulfil the requirements for recoverable packaging when the package becomes waste (Annex II, para. 2).

Specifications regarding recoverable packaging Packaging materials must be fabricated so that a certain percentage by weight can be reused in the creation of new, marketable products, in compliance with EC standards. The percentage of materials to be recovered depends on the type of materials used. Packaging materials which are designed to be composted must be sufficiently biodegradable such that they do not adversely affect the quality of the compost or the composting activity. Packaging identified as "biodegradable" must be able to undergo chemical, biological, thermal or other physical decomposition such that the majority of the finished compost ultimately becomes carbon dioxide, biomass and water (Annex II, para. 3). Furthermore, packaging materials which are designed to be recoverable in the form of energy (including incineration) must have a minimum inferior calorific value which will allow optimisation of energy recovery (Annex II, para. 3(b)). Previously, this figure was set at 13 MJ/kg (the approximate energy value of cardboard and paper), but it has now been left open.

Specifications which were deleted The final version of the Directive deleted an important caveat that incinerated packaging materials could not be substituted for other fuels if they had a greater effect on the environment than the substituted fuel. With the problems caused by the mounting pile of recycled materials, incineration has slowly gained greater acceptance as a form of recovery. The inclusion of incineration provisions in the Directive is a recognition that the use of only reuse and recycling is not sufficient to meet the prescribed targets. Yet, ironically, incineration may now be used, even if it results in greater harm to the environment than the source of energy for which it is being substituted.

Logistical requirements

Two important requirements were added to the final version of the Directive, reflecting the demands of industry that the Directive be more responsive to the realities of the marketplace. The first condition was that Member States cannot apply the requirements of the Directive to packaging which existed prior to 31 December 1994. The second limitation was that until 31 December 1999, a Member State must permit the use in its marketplace of packaging manufactured prior to 31 December 1994, as long as the packaging is in conformity with its national law (Article 22). The realities of the marketplace suggest that there will be 'pre-Directive' packaging on the market for years to come. Manufacturers will be able to refine such packaging by adding the recycling/recovery system symbol used in the Member State where it is marketed. However, as consumers become better educated, they will likely demand the newer packaging which seems more environmentally sensitive.

In addition, each Member States is forbidden from impeding the introduction of packaging materials into its territory if the materials comply with the Directive (Article 18). Yet although several Member States, notably France and Germany, have enacted a procedure for co-operation regarding the use of the Green Dot in their respective markets, firms remain confused by the mass of packaging regulations encountered when marketing a product throughout the Community.

Obligation upon Member States: prevention

Hierarchy for packaging waste and prevention

The establishment of a hierarchy for packaging was highly disputed during the negotiation process. Initially, Belgium led a group of Member States calling for a hierarchy in which a reduction in the production of packaging waste was the first priority. The second priority, reuse, was to be followed by recycling; then recovery with incineration; recovery without incineration; and then landfilling as a last resort (Environment Europe, 1993). However, after several years of debate, the issue of hierarchy was reduced to two paragraphs in the Preamble and Article 4.

The Preamble acknowledges that the best method for reducing packaging waste is to reduce the total amount of packaging produced. The Preamble also states that life-cycle assessments ('LCAs') should be developed so that short of prevention, a hierarchy may be established among reusable, recyclable and recoverable packaging. However, the Directive then makes a further declaration that in light of the present technology regarding LCAs, reuse and recycling should be preferred over other techniques. Industry claims that this aspect is ambiguous and may lead to discrimination against methods such as incineration with energy recovery, noting that as the Directive forbids Member States from discriminating against packaging which meets the criteria of the Directive, packaging which can be incinerated should be equally acceptable under the Directive (EIS, 1995).

Article 4 is rather weak, in that instead of containing 'hard law', it urges Member States to take action to prevent the formation of packaging waste. Member States are forbidden however, from taking measures which would not meet the essential standards of the Directive or would create any obstacles to trade within the Community.

Heavy metals

In contrast with previous drafts of the Directive which listed the permissible concentrations of heavy metals in an annex, the final version of the Directive contains a separate article which provides that the concentration levels of lead, cadmium, mercury and hexavalent chromium present in packaging shall not exceed 600 ppm by weight after 30 June 1998; 250 ppm by weight after 30 June 1999; and 100 ppm by weight after 30 June 2001 (Article 11, para. 1). However, the Directive contains a specific exemption for packaging entirely made of lead crystal glass (Article 11, para. 2). Article 11 further provides that pursuant to the Committee Procedure, the Commission will determine to what extent these provisions do not apply to packaging made from recycled materials. The Commission may also determine to what extent packaging can be exempted if it is part of a closed and controlled chain of use. Lastly, the Commission may exempt certain types of packaging from the 100 ppm requirement (Article 11, para. 3). To date, Directorate General XI of the Commission is still working on these issues.

Obligations upon Member States: information collection and provision

Information systems and information for users of packaging

Articles 12 and 13 require Member States to create information systems for both the Commission and the users of packaging. For example, Member States must create databases on packaging and packaging waste so that the Commission may review the process of implementation. The databases must comply with Annex III, which sets forth the formats to be used for the compilation of databases. The formats have been greatly simplified and reduced from those of prior versions of the Directive, in recognition of the difficulties the Member States will encounter in compiling this type of information. The formats are now designed to elucidate (for primary, secondary, and tertiary packaging) the overall quantities consumed within a Member State, as compared with quantities reused; quantities recovered and disposed of; and quantities recycled or recovered. For the latter two categories, the totals are first divided between household and non-household waste. All charts differentiate among glass, plastic, paper/cardboard, cardboard composite, metal and wood (Article 12, Annex III). Further, Article 12.4 specifically requires Member States to take into consideration the potential problems created for SMEs by the collection of such data. This provision reflects the concern in the Preamble regarding the effect the Directive will have on the financial stability of SMEs.

In February of 1997, the Commission issued Decision 97/138/EC[8] regarding the harmonisation of the database formats. The purpose of the Decision was to clarify and streamline those points of the Directive regarding the presentation of data which were unclear or remained unresolved. In accordance with the Directive's goal of keeping up with technological advances, Decision 97/138/EC adds a definition of 'composite' packaging: that which is 'made of different materials, and which cannot be separated by hand, none exceeding a given percent by weight, which shall be established in accordance with the [Committee Procedure]'. Potential exemptions for some materials may be established by the same procedure'. In addition, Decision 97/138/EC further harmonises the Tables contained in Annex III of the Directive. The distinction between household and non-household waste has been eliminated, and there is now a distinction between municipal and non-municipal waste. However, the only data which Member States must provide is information regarding glass, metals, plastics, paper and fibreboard.

Lastly, Article 4 of Decision 97/138/EC requires that each Member State present a report to the Commission for each five-year period, commencing 30 June 2001, which report shall supply quantitative information regarding packaging waste which is hazardous as a result of product contents (pursuant to Directive 91/689/EEC and Council Decision 94/904/EEC).

The purpose of the databases is to give the Commission an idea of the extent, attributes and development of packaging and packaging waste for each Member State and in particular, issues such as toxicity and types of materials used. Member States must require all economic operators to provide them the data needed to comply with the Directive (Article 12). Unfortunately, the reality is that some Member States simply do not have the resources to assure proper compliance.

With respect to end users, Member States were required to take measures within two years of the Implementation Date, i.e., 30 June 1998, to educate users of packaging (particularly consumers) about the return, collection and recovery systems available. Member States must also advise users as to how they can participate in the recovery, reuse and recycling of packaging materials. Users must also be kept informed of the various environmental markings on packaging 'in the market' (which would include all markings in use in the EC), and of the management plans for packaging and packaging waste (Article 13). These provisions reflect the fact that end users are an essential element of packaging management systems. The Preamble notes that end users need to be advised of the systems for the recovery and recycling of packaging materials and of reusable packaging materials so that they are encouraged to participate in the programme. Ironically, due to the German government's educational efforts, the recovery of waste materials was so successful that this Member State became one of the world's biggest exporters of packaging waste (Cairncross, 1993). As for Member States in general, it is difficult to know what measures (if any) were taken by the 30 June deadline, as the Directive provides the Commission with absolutely no powers with which it can judge compliance. Thus, the education of end users is in a sense a measure left to the initiative of the Member States. However, Question 7 of Decision 97/622/EC, *infra,* does require Member States to indicate what measures they have taken to educate end users.

Management plans

Member States are already required to prepare 'waste management plans' pursuant to the Waste Directive.[9] The Directive requires that Member States now add a separate section regarding management plans with respect to packaging materials, and in particular, address measures taken regarding prevention and reuse (Article 14). In June of 1998, the Commission decided to send Reasoned Opinions to Belgium, France, Greece, Luxembourg, Ireland, Italy and the Netherlands for failure to submit waste plans for packaging, hazardous or general waste, as required by several pieces of legislation, including the Directive (DGXI Press Release, 1997a). However, since that time, no further action has been taken, as the Commission is attempting to resolve the issues via informal dialogue.

Obligation to report

Pursuant to Articles 12 and 17, beginning with an initial period from 1995 to 1997 and continuing every third year thereafter, each Member State must submit a report to the Commission in compliance with the Environmental Implementation Directive.[10] To assist with the preparation of this report, as well as others required for various waste directives, in May of 1997, the Commission issued Decision 97/622/EC,[11] regarding the implementation of certain waste directives. Decision 97/622/EC contains a questionnaire to be used as a basis when preparing the Article 17 report. To date, no reports have been filed for the initial period of 1995 to 1997, due to the decision to wait until the Commission had prepared the questionnaires. Indeed, the Preamble to Decision 97/622/EC states that, 'the first sectoral report will cover the period 1998-2000, inclusive'.

Obligations upon the institutions of the Community

Marking of the packaging materials

The article on markings was sharply reduced in the final version of the Directive, reflecting the fact that Member States could not reach agreement on what marking system should be used. Member States currently use varying symbols to identify packaging which is recyclable; reusable; or made from recycled materials. The final version required that no later than 31 December 1996 (two years after the Directive entered into force), the Council would have to formally decide on the system to be used (Article 8). On 25 November 1996, the Commission set out a proposal for markings, which was then to be subjected to the co-decision procedure. Although more than two years have elapsed since the proposal was first presented, it has become mired in controversy.

Conflicts have arisen because while in Germany and France, the Green Dot symbol of the two-toned inter-wrapped arrows is used on packages eligible for the national systems, in other Member States, treatment systems use the international 'chasing tri-arrows' symbol to mark qualifying packaging. Still other programmes (in Italy, for example) have their own symbol. While the selection of the 'European symbol' from among the choices may not seem substantial, at stake here is the age-old problem of national pride and influence, complicated by the fact that within Europe, the Green Dot symbol is now fairly well-known. Furthering the problem is the idea that in order to clarify the situation for the consumer, one symbol should be used for packaging which is eligible for the national packaging waste treatment system, while another symbol should be used to identify packaging which is made from recycled materials.

In addition to the need for the above markings, it was clear that an identification system was necessary to assure the correct treatment of packaging waste. To that end, the Commission was mandated to propose by December 1995 a numbering and classification system as to the nature of materials used in the packaging (Article 8.2). The Directive further requires that manufacturers place the composition information on either the label or on the packaging; information must be sufficiently visible, legible, and durable such that it will last through the consumer's use of the package, including after the package has been opened (Article 8.3). Article 8.2 makes reference to Annex I, which specifies the following numbering system: 1-19 for plastic; 20-39 for paper and cardboard; 40-49 for metal; 50-59 for wood; 60-69 for textiles; and 70-79 for glass.

The development of the numbering and classification system turned out to be one of the most contested aspects of the Directive. Beginning in the late 1980s, industry developed its own complex numbering system whereby each package is stamped with a number and barcode. A scanner can 'read' the package and determine its composition, including whether it is made from recycled or virgin materials. This system is used throughout the world and thus facilitates the export of EC products. Industry remains concerned that the Community institutions will try to establish a new set of markings. Packaging manufacturers argue that it is unnecessary and redundant to pay for the creation of a new system. Further, the changeover to a new classification system would have to be completed at great cost to the packaging industry (EIS, 1995). Indeed, it is more rational to give the current system a chance to work before scrapping it for a new, untested method. In the end, in February of 1997, the Commission issued Decision 97/129/EC[12] which resolved this issue. Decision 97/29/EC keeps essentially the same marking system as in Annex I of the Directive, but adds an additional section regarding the numbering and abbreviation system for composites.

Standardisation

The Commission is required to promote European standards as to the essential requirements of Annex II, i.e., the composition and the reusable, recoverable and recyclable nature of packaging materials.[13] The Commission has mandated the European Committee for Standardisation ('CEN') to prepare these standards. Similar to what transpired with the drafting of the Directive, CEN must develop standards which achieve the dual goals of protecting the environment and facilitating the functioning of the internal market. The standards, which will probably not be ready before 2000, will be developed with the participation of trade and industry, scientific bodies, consumer groups, and environmental interest groups (EIS, 1996c). The inclusion of the latter two groups is particularly important for establishing transparency in the development process. Interestingly, the opinion of industry is that despite the input of 'green groups', these standards are not likely to establish new ground, but instead will reflect the current situation. This is due to the fact that during the last decade, firms have improved their technologies so as to make them more environmentally friendly. However, this 'greening' of industry has largely come about via pressure from consumer groups and environmentalists. Thus, the input from these groups may indeed influence the standards more than industry might think.

The Committee Procedure

Article 21 outlines a special procedure by which the Directive may be amended. The Committee Procedure is necessary because the packaging industry is continually researching recovery and recycling technologies and because it is not yet clear what effects European-wide packaging legislation will have on industry. Via the Committee Procedure, the Commission receives assistance from a committee of representatives from the Member States and a Chairperson from the Commission. The Chairperson submits a draft of a Directive revision to the committee, which delivers its opinion within a specified time frame; the length of time to render an opinion is determined by the Chairperson in response to the urgency of the matter. The decision of the committee is determined by qualified majority, pursuant to Article 148 of the Treaty.

Thereafter, if the committee approves the revision, it is adopted by the Commission. If the committee does not approve the revision or fails to deliver an opinion, the Commission is required to submit a proposal of its own directly to the Council, which thereafter has three months in which to make a decision. If the Council fails to act within three months, the Commission may adopt the revision (Article 21). This article is quite a change from the previous drafts, in which the committee's authority was merely advisory. Clearly, the Member States wished to retain greater control over the implementation of this particularly sensitive directive. Although the committee has met since the Implementation Date, no special derogations have yet been given.

Options left to the Member States

The Directive contains several permissive provisions, which allow Member States some leeway in their implementation. Two of the issues, reuse systems of packaging and economic instruments, have proved so complex and/or controversial, that they are the subject of separate chapters of this book. What follows is a brief overview of the provisions of the Directive that directly address these issues.

Reuse systems for packaging

The final version of the Directive deleted previous definitions of 'reusable packaging', 'non-returnable packaging', and 'one-way packaging' and substituted for them an expanded definition of 'reuse':

> any operation by which packaging, which has been conceived and designed to accomplish within its life cycle a minimum number of trips or rotations, is refilled or reused for the same purpose for which it was conceived, *with or without* the support of auxiliary products present on the market enabling the packaging to be refilled; such packaging will become packaging waste when no longer subject to reuse (Article 3.5, emphasis added).

Thus, a package is considered *reusable* as long as it can be refilled and/or reused for its original purpose. When the package can no longer fulfil its purpose, it becomes packaging waste which may then be *recycled* and/or *recovered*.

Article 5 is a permissive rather than prescriptive article: it states that 'Member States *may* encourage reuse systems of packaging, which can be reused in an environmentally sound manner, in conformity with the Treaty' (emphasis added). Several Member States have relied on Article 5 as a basis for establishing minimum quotas for packaging which is reusable or refillable; yet, these quotas are detrimental to packaging which is recyclable.

The interpretation of this article has thus led to a schism of opinion among the Member States and within the Commission itself (Long et al., 1997).

Economic instruments

While most Member States have set up systems similar to those of *Duales System Deutschland* or *Eco-Emballages,* some have additionally created an eco-tax system. The Preamble notes that Member States may use 'economic instruments' in accordance with the Treaty, in order to avoid new forms of protectionism. Article 15 sets forth the applicable law; this article engendered extensive controversy during the negotiation of the Directive. The final version of the Directive states that, '*....the Council* adopts economic instruments to promote the implementation of the objectives set by this Directive' (Article 15, emphasis added). In the absence of such measures, Member States may adopt legislation designed to implement the Directive's objectives; however, Member States must act 'in accordance with the principles governing Community environmental policy, *inter alia,* the polluter-pays principle, and the obligations arising out of the Treaty'.

The final version of Article 15 is quite a change from prior drafts which had delineated the restraints to be placed on the use of economic instruments by Member States. Such restraints had included the principal of proportionality and the requirement of a clear link between the measure and its goal. Member States had also been required to apply the measure without discrimination among economic operators or among the various categories and types of packaging. Implementation of the measure was further required to be straightforward and could not demand excessive administration. Lastly, the measure could not impede tax legislation. But, while at first glance it may seem that the Member States have achieved a victory with the final version of the Directive, the deleted provisions largely reflect the case law of the European Court of Justice and thus, an economic instrument would be subject to those principles in any event.

Although it might appear that economic instruments must be notified to the Commission pursuant to Article 16, this article excludes measurements of a fiscal nature. The Amended Proposal had added a subsection which would have required Member States to notify the Commission of all economic instruments implemented for the purpose of complying with the Directive. In the event the economic instrument involved the distribution of state aids, the Member State would also have been required to notify the Commission in advance of the distribution. Yet, although these provisions were deleted, the Member States remain subject to the state aid provisions of Article 92 et. seq. In this regard, the Community institutions retain some control over the Member States.

Voluntary agreements

The final version of the Directive added a definition of voluntary agreements, in recognition of the Dutch, Danish and Belgian packaging agreements then in existence. In 1995, Finland also executed an agreement regarding packaging waste and the Netherlands revised its Packaging Covenant in 1997. The definition applies to any formal agreement negotiated between a Member State and the economic sector concerned. The Directive requires that the formal agreement be accessible to any entity which agrees to meet its conditions in respect of the purpose of the Directive (Article 3.12).

Economic operators

The definition of 'economic operators' includes not only packaging producers and converters, traders and distributors, and suppliers of packaging materials, but also importers of packaging materials and statutory organisations and authorities which are affected by the packaging process (Article 3.11). This definition makes the Directive applicable to all entities which have any relation to the packaging process, including organisations such as *Eco-Emballages* in France, *Duales System Deutschland* in Germany and *Valpack* in the United Kingdom, thus broadening the scope of the Directive.

Post-enactment obligations of the Directive

Implementation Date

Member States were required to transpose the Directive into national law within 18 months of the Directive's entry into force on 31 December 1994, i.e., before 30 June 1996. The Directive required national authorities to immediately notify the Commission thereof, as well as any laws, regulations and administrative procedures which they might adopt within the scope of the Directive. Any such measures in national law are required to contain a reference to the Directive (Article 22). As of the Implementation Date, most of the Member States had not yet notified the Commission (Long et al., 1997).

Thereafter, for the next two years, the Commission proceeded to notify reasoned opinions and to send letters of formal notice to several Member States (notably, Finland, Belgium, Portugal, the United Kingdom and France) for their alleged failure to properly transpose all provisions of the Directive.

In June of 1998, the Commission even made applications to the European Court of Justice against Luxembourg, Ireland and Greece for failure of transposition (DGXI Press, 1997b, 1998a,b,c). As normally happens at this point, discussions began between the Commission and the Member States concerned; by September of 1998, it was considered that all Member States had notified the Commission of their implementing legislation.[14]

Yet, the formal notification of implementing legislation is only the beginning of the process, as the Commission must transmit the proposed legislation to the other Member States for their review and comment, as well as prepare an analysis of its own. In fact, as of July 1997, the Commission had adopted nine detailed opinions regarding fourteen notifications (Long et al., 1997), which suggests the high level of contention regarding the interpretation of the Directive.

Following a few of the notifications, several issues were raised by the detailed opinions. Luxembourg had originally proposed an eco-tax as well as quotas on refillable containers for liquids. But after concerns were expressed by other Member States via detailed opinions, Luxembourg finally dropped these provisions and is now considered to have correctly ratified and transposed the Directive.

Denmark, on the other hand, continues to insist that its legislation banning metal cans does not infringe either the Treaty or the Directive. Although in July of 1997, the Commission announced that it would commence infringement proceedings, it has only recently begun Article 169 proceedings. As is normal, the Commission continues to attempt to resolve the issue through dialogue with Denmark.

Lastly, the situation with Germany's transposition remains unresolved. Germany's situation has of course, always been different, due to its experience with the German Packaging Ordinance, which was enacted three and a half years before the Directive. Unlike some other Member States, which were able to simply recopy the provisions of the Directive into national law, Germany had to merge the two pieces of legislation. Since Germany first notified its implementing legislation, it has worked with the Commission to resolve most of the disputed issues, in particular, those regarding proper transposition of Article 9 and the provisions of Annex II (heavy metals). Germany's modification of its legislation was approved in 1998 by the *Bundesrat*. However, Germany has refused to modify any legislation which contains quotas for refillable/reusable containers. Thus, the new legislation still contains a quota of 72% for non-milk liquid containers and 17% for milk containers. Germany has however, eliminated separate quotas for the Länder.

Notification of framework measures

Article 16 states that before adopting a measure which falls within the framework of the Directive, Member States are obliged to submit drafts to the Commission. Measures of a fiscal nature are excluded, except to the extent they contain or are meant to encourage compliance with, technical specifications. When the proposed measure also concerns a technical matter, the Member States are permitted to submit a simultaneous notification for Directive 83/189/EEC. The rationale for this article is to allow the Commission and other Member States to comment on the possible impact of such measures on the market, and to object as necessary (Long et al., 1997). Similarly, Article 9 requires Member States to notify the Commission of the text of those national standards which they believe comply with the Essential Requirements of the Directive. This notification is also subject to the procedure of Directive 83/189/EEC.

Directive 83/189/EEC provides that Member States are required to transmit to the Commission the draft of any technical regulation, unless the draft merely transposes the complete text of an international or European standard (in this case, general information regarding the standard is enough). If the draft does not make it clear, the Member State must also indicate the grounds on which the draft is based. Any later changes to the draft must be communicated as well. The Commission then notifies all other Member States of the draft and thereafter, the Commission and Member States have three months in which to make comments. If there is a complaint that the legislation will create obstacles in the internal market, the Member State concerned must wait six months for further review. And, if the Commission announces that either: 1) it intends to propose or adopt a Directive on a similar matter; or 2) there already exists Community legislation on this matter, the Member State must wait twelve months for further review.

After receiving the comments, the Member State must indicate to the Commission what action it intends to take in response. The Commission must thereafter comment on the reaction. At the end of the procedure, the Member State concerned must transmit the definitive version of the legislation to the Commission. Although the full procedure is long and complicated, disputes are usually resolved via dialogue with the Commission.

The importance of a Member State complying with the above procedure was underscored by the 1994 case, *CIA Security,*[15] in which the European Court of Justice ruled that a national court cannot apply the provisions of a national technical regulation which has not been notified in conformity with Directive 83/189/EEC. Moreover, any Member State which does not notify such legislation is open to an infringement action by the Commission.

Criticisms of the Packaging Waste Directive

The Directive has been criticised for its failure to address the consequences of such a far-reaching environmental programme. For example, the Directive does not identify measures to stimulate the demand for recycled products; nor does it set requirements which specify a minimum content of recycled materials in new products. Further, the Directive does not set minimum standards with regard to public procurement of recycled products or products made with recycled materials; such measures can be significant because they set an example for the Community.

In addition, the Directive fails to address the future impact of the legislation on the packaging marketplace. Such effect is surely not difficult to envision, as the economic effects of Germany's *Duales System Deutschland* have been felt for several years (for example, when the scrap paper market in the United Kingdom collapsed). And while the scrap paper market itself has rebounded (EIS, 1995; 1997b), it is not the only market to suffer from over-capacity. It has been estimated that the Directive's recycling levels of 60% could lead to a recycled materials market which is as large as the virgin materials market (Dallemagne et al., 1993). Thus far, the only acknowledgement of this problem has been in the Preamble, which notes that to achieve the recycling objectives while avoiding distortion of trade and competition, markets for recycled packaging materials must be both developed and expanded.

Lastly, the Directive fails to address the needs of regional and local cultures. Stories abound of situations in which a collection and recovery system had to be abandoned because it simply was not economically feasible. If the Directive is to be successful, it must make specific provisions for regional variations in both packaging needs and recycling abilities. An example of the difference is the returnable bottle issue. The success of such a programme is dependent upon the availability of sufficient fresh water to clean the bottles, the density of the population in a particular area and the transportation systems (Dallemagne et al. 1993). Thus, while a collect-and-refill bottle system might be successful in Amsterdam, it would probably not be successful in southern Spain.

On the other hand, the Directive has spurred technological development that probably would not otherwise have occurred. The Directive has also forced companies to rethink the entire package of their product, sometimes resulting in cost reduction. Moreover, the Directive has led to a heightened consumer awareness of the effect of packaging waste on the environment and to greater efforts to recycle and reuse packaging. Of course, the level of these efforts is determined in part by how well the Directive is implemented in each Member State. And as Chapter 3 suggests, the extent of implementation can vary substantially.

Notes

1 European Parliament and Council Directive 94/62/EC of 20 December 1994 on packaging and packaging waste, *Official Journal* No. L 365, 31 December 1994.
2 Proposal for a Council Directive on packaging and packaging waste, *Official Journal* No. C 263, 12 October 1992.
3 Amended proposal for a Council Directive on packaging and packaging waste, *Official Journal* No. C 285, 21 October 1993.
4 Directive 75/442/EEC of 25 July 1975 on Waste, as amended by Directive 91/156/EEC *(Official Journal* No. L 78/32, 18 March 1991). The categories are further detailed in Appendix B.
5 Annex II B of the Waste Directive is detailed in Appendix B.
6 'The countries of the north had to put a lot of water in their wine'. ('Emballages Européennes', *Emballages*, February 1994, p. 42.)
7 For an discussion of the procedure required by this directive, see Notification, *infra.*
8 Commission Decision 97/138/EC of 3 February 1997 establishing the formats relating to the database system pursuant to European Parliament and Council Directive 94/62/EC on packaging and packaging waste, *Official Journal,* No. L 52, 22 February 1997, p. 22 (see Appendix C).
9 See footnote 4.
10 Directive 91/692/EEC, standardizing and rationalizing reports on the implementation of certain directives relating to the environment, *Official Journal,* No. L 377, 23 December 1991, p. 48.
11 Commission Decision 97/622/EC of 27 May 1997 concerning questionnaires for Member States reports on the implementation of certain Directives in the waste sector (Implementation of Council Directive 91/662/EEC), *Official Journal*, No. L 265, 19 September 1997, p. 13 (see Appendix C).
12 Commission Decision 97/129/EC of 28 January 1997 establishing the identification system for packaging materials pursuant to European Parliament and Council Directive 94/62/EC on packaging and packaging waste, *Official Journal*, No. L 50, 20 February 1997, p. 28 (see Appendix C).
13 Standards are also to be developed for criteria as to the following packaging issues:

- Criteria and methodologies for life-cycle analysis of packaging;

- The methods for measuring and verifying the presence of heavy metals and other dangerous substances in the packaging and their release into the environment from packaging and packaging waste;
- Criteria for a minimum content of recycled material in packaging for appropriate types of packaging;
- Criteria for recycling methods;
- Criteria for organic recovery (composting methods); and
- Criteria for the marking of packaging (Article 10).

14 Although Greece has not yet made a formal notification, it is generally considered to have complied in this respect.

15 CIA Security International SA *v* Signalson SA and Securitel SPRL, Case C-194/94, [1994] ECR-2201.

3 Packaging Legislation of the Member States

Introduction

Among the Member States, there exists a broad spectrum of packaging materials laws, from countries such as Greece, which has recently drafted its first packaging law, to Germany, which enacted a controversial and complex packaging ordinance. When the time came to draft the Directive, as so often happens with EC legislation, the Commission became the centre of a political war among the Member States and industry as to which of their packaging laws was the 'best'. As discussed in Chapter 2, the final version of the Directive is a compromise in the form of a framework directive which permits various types of packaging legislation. Ironically, due to the cultural diversity of Europe, the composition of packaging materials waste differs within the regions of the EC and as a result, a packaging materials scheme which is 'best' for all Member States probably does not exist (Sinclair, 1993).

Article 20 of the Directive required all Member States to enact all 'laws, regulations, and administrative provisions necessary to comply with the Directive' (and to notify the Commission thereof) before 30 June 1996. However, only about half of the Member States notified the Commission by 30 June (and thereafter began discussions regarding their transposition of the Directive into national law). As of July 1997, only half of the Member State had received approval of their drafts and had completed implementation. As it has turned out, the Commission and the Member States have highly divergent views as to what legislation satisfies the requirements of the Directive (Long et al., 1997). Not surprisingly, particularly as this is a framework directive, different packaging regulations have now been implemented in the Community, in terms of transposition and the authorities to be involved in implementation. These differences are rather ironic, given that the idea behind the Directive was to avoid a proliferation of divergent legislation among the Member States, much as had already begun to exist prior to the passage of the Directive.

This chapter begins with an overview of packaging waste legislation in the various Member States, prior to the enactment of the Directive; the implementing legislation of the Member States is then discussed. Rather than analysing the situation on a Member State by Member State basis, Member States are grouped according to common denominators, such as the collection and sorting system used and the establishment of similar targets.

Packaging waste legislation in the Member States prior to the Directive

The Liquid Foodstuffs Directive and environmental agreements

Prior to the enactment of the Packaging Directive in 1994, the level of packaging legislation of the Member States varied from substantial to none at all. In implementation of the Liquid Foodstuffs Directive of 1985,[1] several Member States (including Belgium, France, the Netherlands and the United Kingdom) had concluded environmental agreements with industry. The validity of several of these agreements were questioned for their absence of legal foundation; the closed nature of their negotiation; the difficulty in imposing sanctions to enforce them; and the inability of affected third parties to contest their validity before a court. Moreover, environmentalists criticised the agreements as not being sufficiently focused on prevention.

Ultimately, the Commission brought an action against France,[2] on the grounds that France had failed to comply with the Liquid Foodstuffs Directive by not achieving the targets therein. The Commission argued that it was not sufficient implementation of the directive to delegate the process to the private sector via an environmental agreement, without including more specific obligations on the part of the state. The ECJ disagreed with the Commission, ruling that an environmental agreement would be an appropriate method for implementing the directive, but also holding that the agreement used by France was insufficient because it contained neither specific objectives for the reduction of packaging waste, nor provisions to assure the periodic review of the objectives, as required by the Liquid Foodstuffs Directive.

Other types of pre-Directive packaging legislation

In addition to environmental agreements, Member States used various other types of legislation to address this issue of packaging waste prior to the

Directive. This section groups the Member States chronologically and discusses their situation at the time of the enactment of the Directive.

Denmark

Denmark has been the forerunner of the Community regarding the adoption of packaging materials regulations. By 1971, beer and soft drinks had to be sold in reusable glass bottles which carried a deposit and by 1977, non-refillable soft drink bottles were totally banned. In 1981, a ban was issued on non-refillable containers for beer and soft drinks, which effectively meant that cans could not be used. Following intervention by the Commission, Denmark permitted the sale of beer and soft drinks in non-approved containers, but only in limited quantities. In 1981, Denmark created a deposit scheme for non-refillable beer and soft drink containers which was partially upheld in the *Danish Bottles* case.[3] In 1986, the collection and sorting of household and commercial waste began, as well as the labelling of recycled packaging. From 1988-1990, taxes on packaging and the deposit programme were extended, and taxes on all non-reusable beverage containers were introduced. By 1991, the government had reached agreement with the plastics industry regarding the use of PVC in packaging and had unveiled its waste management programme (Hempen et al., 1993).

Germany

Although Germany had debated various forms of packaging legislation since the 1970s, it was not until 1988 that it enacted its first packaging-specific law, the Ordinance on Drink Containers made from Plastic (the 'Plastic Ordinance'). The Plastic Ordinance imposed both deposits and take-back requirements; it was enacted in response to both the failure of a 1977 environmental agreement on packaging waste and to the announcement of Coca-Cola that it intended to introduce one-way PET bottles. The Commission intervened, opining that the Plastic Ordinance would create trade barriers unless it was enlarged to include all types of beverage containers. Ultimately, the Plastic Ordinance was considered a success, as Coca-Cola decided to use refillable PET bottles instead (Haverland, 1998).

The public's positive response to the Plastic Ordinance led to the German packaging ordinance of 1991 (the 'Ordinance'), which required that packaging be produced from 'environmentally sound' materials (Joly, 1996). The Ordinance's requirement that the volume and the weight of packaging materials be reduced to the bare minimum required for protection and marketing of the goods has resulted in an effort by packaging manufacturers to design more environmentally-friendly packaging, which has ultimately benefited the entire European Community. The Ordinance was notable in that it required producers of packaging materials to reclaim the materials for reuse and/or recycling, thus leading to the creation of 'collection centres' at point of sale. The Ordinance further required that the packaging materials be reusable/refillable if technologically feasible; if not, the materials must be recycled as the ability to incinerate packaging waste was not provided (Hempen et al., 1993). The lack of incineration created extensive problems in Germany, particularly with plastic waste, and eventually led to the acceptance of thermal incineration (Joly, 1996).

Finland

Like Denmark, Finland was an early proponent of packaging waste management, having made plans for the management of both domestic and industrial waste since 1979. However, the first legislation to specifically address packaging waste was the 1990 Law on the Taxation of Packaging, which imposed a tax on all beer and soft drinks which were not packaged with materials which the government had designated for collection and recycling. Finland used a system of deposits and taxes because recycling is not practical in many areas due to great distances involved. Accordingly, the Finnish government has tried to recycle secondary packaging at the point of sale. The pre-Directive regulations mainly involved the collection of paper for recycling and the recycling of beer and soft drink containers.

Sweden

Sweden's first legislation to address packaging materials was its Waste Act of 1990, which required that the producer of any type of waste (specifically including packaging) be responsible for its ultimate disposal. The Waste Act also eliminated one-way cans and bottles and created a taxation system for packaging. In 1993, Sweden passed the Ecocycle Bill, which incorporated the concept of 'Producer Responsibility', that each producer (or importer) should be held responsible for the disposal of a product (including its packaging) in a manner which is friendly to the environment (de Sadeleer, 1995a).

The Netherlands

The Netherlands began its first efforts towards packaging waste management in 1991 with the Dutch Packaging Covenant ('the Covenant'). The Covenant was the result of co-operation between the Ministry of the Environment and the *Stichting Verpakking en Milieu* (Foundation for Packaging and the Environment). The Covenant was a voluntary scheme theoretically binding on its signatories (the 'Packaging Chain'), of which there were more than

500. While the membership was voluntary, over 80% of the packaging industry was involved, due to pressure from both the *Stichting Verpakking en Milieu* and the government. The Covenant took a strong stand on landfilling, prescribing that it must be totally eliminated by the year 2000, and that the Packaging Chain must take back at least 90% of the used packaging which is collected separately and cannot be reused (Hempen et al., 1993).

France

France's packaging waste management began in January of 1993, when the Packaging Waste Decree (the 'Decree') went into effect. The Decree reflected a joint effort of industry following its observations of German industry's problems upon implementation of the German Ordinance. The Decree was the result of a concerted effort between the French Environmental Minister and the presidents of the two largest packaged goods companies in France, St. Ouen and BSN. French industry feared that if it did not become actively involved with the drafting of the Decree, the law would not be workable (Emballages, 1992b). However, the involvement of these tow major firms in the drafting of the law later raised complaints from small businesses (Theiffry et al., 1993). Unlike the initial German system, France permitted incineration as an acceptable method of recovery. Along with Sweden, France was also one of the first Member States to introduce the concept of Producer Responsibility, pursuant to which, if the producer or importer cannot be identified, then the party placing the product on the market is responsible.

Austria

In October of 1993, prior to its entry into the Community, Austria enacted a packaging ordinance which was drafted to comply with the then-proposed version of the Directive. Yet, even prior to the enactment of the Directive, Austria had begun to aggressively recover packaging waste: between 1991 and 1994, the amount of landfilled waste was reduced fifty percent.

Belgium

When it did not have much success with its packaging waste tax, the Belgian government decided to create a recovery system similar to that of France. The new system, known as *Fost Plus,* began operations in March 1994. *Fost Plus* combined the three former pilot programme of Flanders, Wallonia and Brussels, and for the first time in Belgium, reflected the pro-active involvement of the business community (which as experience has shown, is critical to the success of this type of programme); indeed, it was guaranteed that no additional system of taxes or other form of sanctions would be levied against the participants of *Fost Plus.*

Spain

Prior to 1996, the only Spanish legislation regarding packaging had been regulations implementing the Directive of 1985. As other Member States began discussing packaging legislation in the early 1990s, similar debates occurred in Spain. The original Spanish packaging legislation not only encouraged recycling through the prohibition of non-returnable bottles, but also planned to charge fees for the disposal of items such as tires and batteries. However, despite numerous discussions, these proposals never came to pass. Instead, the general attitude was to participate in the debate of the Directive and then adopt its standards.

United Kingdom

In the 1980s, the United Kingdom was highly innovative in the development of disposable plastic packaging in Europe and began substituting plastic in place of glass and metal. At that time, convenience (in the form of lightweight packaging) and not protection of the environment, became the priority for the consumer. In the late 1980s however, priorities began to change as the public became sensitised to landfill issues. In 1990, the British government enacted the Environmental Protection Act, which contained restrictions on waste disposal. While the Act did regulate household waste, it did not specifically address packaging waste (Hempen et al., 1993). *This Common Inheritance*, a 1990 White Paper on the environment, set the goal to recycle and/or compost half of all recyclable household waste by the year 2000. However, industry response to the demands of government were minimal. In the end, the Consortium of the Packaging Chain was created; members include suppliers of raw materials, retailers, recyclers, manufacturers of packaging and of fillers, and waste disposal contractors.

Other Member States

Some Member States had packaging legislation prior to the Directive, but the legislation was very weak or poorly enforced. For example, Italy enacted Decree Law No. 397 in 1988. This decree created collection systems for different types of recyclable packaging and imposed levies on packaging materials; however, it was loosely enforced. In other Member States, the decision was made to watch the results and public reaction, first to the German Ordinance, and then to the proposals for the directive itself. This path was chosen by Luxembourg which, although it did proposed beverage container legislation, chose to wait for the final version of the Directive before going any further.

Irish activity in packaging waste management was rather limited, although in response to the 1985 Directive on Containers of Liquids for Human Consumption, Dublin enacted a bottle collection scheme as well as a programme for a deposit and return system. The Irish government also proposed banning the sale of alcohol and other drinks in metal cans and restricting the use of PET bottles. Originally, the government adopted a strict policy in which taxes or deposits would be levied on various types of packaging deemed harmful. However, this policy was later minimised following the response of industry, in order to encourage increased the use of recycled materials. In Greece and Portugal, it was decided to wait until the Directive was enacted before proceeding further (Hempen et al., 1993).

Member States using a dual system of return and recovery

Germany is the only Member State which has implemented a dual system of return and recovery, i.e., in addition to the traditional municipal waste collection, *Duales System Deutschland* co-ordinates a second collection of recoverable materials. Bags and boxes of different colour combinations (depending upon the municipality) are used to facilitate both kerbside collection (for plastic, metal and paper) and transport of glass by householders to collection centres (ERRA, 1998). This system was created in response to the 1991 German Ordinance, which required retailers to make provisions so that all packaging waste could be taken back at the point of sale.[4] At the same time, a deposit system was implemented on disposable drink containers and packaging for detergents and cleaning materials so that consumers would have a motive to return the packaging after use. Unfortunately, *DSD's* collection efforts wildly exceeded expectations. As of October of 1993, more than 100,000 tons of packaging materials were being stored throughout Germany and approximately 95% of the German households were participating in the recycling programme (Jackson, 1993).

In the first few years of the programme, *DSD* incurred a debt of more than five hundred million ECUs. The debt was caused by several problems: first, *DSD* had to pay local governments for the collection and storage of the backlog of uncollected materials; second, *DSD* failed to collect funds from many of the participating companies, some of which had severe financial problems themselves. In addition, *DSD* suffered from a free rider problem in that it ended up treating the packaging waste of all firms, not just that of undertakings that had paid for the Green Dot. Upon concluding that the programme was a failure, several German cities threatened to withdraw from the system.

While the governmental participants of *DSD* may have been angry, the private members of *DSD* were even angrier. A Bavarian dairy company, Müller-Milch, challenged the Ordinance as unconstitutional and tried to withdraw from the programme, arguing that it was being forced to subsidise a failing system.[5] The Court opinion, which was upheld on appeal, denied Müller-Milch the right to withdraw, ruling that there exists in Germany a strong public interest of environmental protection, which interest far outweighs Müller-Milch's interest in saving money. Moreover, the Court noted that as *DSD* was in its infancy, it needed the support and protection of the German government. After this judgement, many companies decided to accept the situation and several loaned funds to *DSD* to cover its shortfall. Due to the support of the German government and German industry, as well as decreased storage costs after it was able to incinerate its plastic waste, *DSD* was financially solvent by 1995.

The effect of the Ordinance has been wide-spread, partly because it is the first environmental law to affect all undertakings, in a manner which is similar to that of a company income tax. In contrast, previous environmental regulations had focused on a particular industry or on a particular resource. As a result of its far-flung effects and of the information campaign produced by the government, the Ordinance has forever changed the way German consumers and the retail shopkeepers view packaging. Its effects have led to the development of many new, 'environmentally friendly' products and packaging, which are now being sold throughout the Community. While the Ordinance has been criticised for failing to effectuate much change, in terms of the amount of packaging produced or its composition (Joly, 1996), it *is* doubtful that the excessive packaging of the past will ever return.

Member States using a single system of return and recovery

In contrast to the German system, the rest of the Member States have tended to follow France's return and recovery system, which was designed after it conducted a study of *DSD*. The French system involves a single collection of all waste by municipal authorities; thereafter, the waste is sorted at regional centres before being recycled, recovered or incinerated (often at the same facility), or shipped for landfilling. Although the French system (or a variation of it) has been copied by other Member States, it is unclear as to whether the attraction of the French system was its perceived environmental benefits or its financial advantages, as the system was touted early on as being much cheaper than that of *DSD*. Interestingly, *DSD* initially insisted that its greater costs were caused by the storage of plastic waste that could not be incinerated in Germany; now that it has been incinerating plastics for several years, *DSD's* per tonne cost of approximately 400 Euros is still the highest in the Community, more than 100 Euros greater than the cost of Austria's ARA, the second most expensive system (Price, 1998).

While the French system is more 'environmentally friendly' in that the diminished transport results in reduced emissions and road tear, it does have its drawbacks in terms of worker health and safety. For, while having the sorting done by workers instead of householders does create new jobs (and thus is politically popular), the work can be dangerous (due to the conditions of the packaging after use) and unsanitary (due to the residues of foodstuffs and chemicals). On the other hand, the French system is viewed as more flexible, as it allows administrators to concentrate on those areas where the recovery and recycling targets of the Directive can more easily achieved, i.e., metropolitan areas. Accordingly, in some Member States, waste from remote areas is not subject to any type of system, as the cost (both financially and to the environment) is judge to exceed the benefits.

In implementing the single system of return and recovery, most Member States have one umbrella organisation that is mandated to supervise the entire process. This organisation then sub-contracts with smaller associations, each of which is usually specialise in one material, i.e., glass or paper. Each sub-contractor (which is usually closely allied with that industry) then arranges for the collection, sorting and treatment of its particular material. Each sub-contractor is also responsible to the umbrella organisation for meeting the targets of the Member State legislation for its particular material. This type of programme may be found, for example, in Austria, France and Italy. Yet, while most Member States have implemented similar systems of return and recovery, there are unique features of several of the Member States systems.

For example, the United Kingdom and Austria have special exemptions from their legislation for small producers/importers of packaging, with criteria depending upon such factors as weight, volume or annual turnover. Finland, on the other hand, requires that producers who annually sell in excess of a certain amount of packaging must register with the Packaging System Producer Register. Belgium, which has always been focused on prevention, requires that every three years, each producer/importer who places more than 10 tonnes of packaging on the Belgian market must prepare a packaging prevention plan. Several Member States (France, Belgium and Portugal) have set up packaging councils to work with both industry and NGOs to address such issues as packaging prevention and encouragement of reuse/recycling by the public (ERRA, 1998).

With regard to financial issues, two Member States have taken differing approaches. Spanish legislation contains a provision that should industry not meet the Directive's minimum target of recycling 15% of each material, eco-taxes be summarily imposed. At the other end of the spectrum, SVM.PACT, the packaging association of the Netherlands, has vowed to have minimalist presence, refusing to impose a financing system, but instead relying on the market principles of Covenant II. SVM.PACT will thus only charge a subscription fee, which will be used for its reduced costs of administration and monitoring (ERRA, 1998).

Member States eligible for special deferments of the targets

Although the Directive grants three Member States, Greece, Ireland and Portugal, an extension of time in which to meet its targets, all three have been active in setting up systems and meeting their goals. In Greece, the government has turned to industry to shoulder much of the responsibility. The Hellenic Recovery and Recycling Association ('HERRA') has taken the lead in establishing return and recovery systems and believes it will achieve the mandated 25% recovery rate in the year 2000. HERRA views its biggest problem as the lack of information available to the public (ERRA, 1998). One should note however, that the Directive provides that information dissemination is the responsibility of each Member State and not the national associations.

Ireland has worked hard to implement its systems; it should easily meet the 25% target by 2000. Ireland has however, two serious problems: that of free riders on its collection schemes and that of a lack of recycling facilities and markets for secondary materials. Portugal has similarly been moving forward: it should also meet the 25% recovery rate and has even implemented a joint recovery programme which uses the Green Dot.

Thus, despite the initial worries at the time of the enactment of the Directive, these three Member States seem to be moving along at a good pace. However, whether they will be able to meet the final targets of the Directive by the deferred date of 31 December 2005 remains to be seen. As was suspected, the biggest problems for these Member States appears to be the lack of markets for the recycled materials, which has in turn discouraged investment in this area.

Allocation of obligations along the packaging chain

Member States have chosen different methods of allocating responsibility for packaging waste along the packaging chain. One may distinguish between 'multi-point' and 'single-point' obligations. The multi-point obligation, which is the most common, places responsibility for packaging waste along the length of the packaging chain; accordingly, all members of the packaging chain must work together to achieve the targets of the Directive. This obligation may be found, for example, in Austria and Finland (Reid, 1995).

In contrast, a single-point obligation places responsibility for packaging waste on only one point in the chain, usually on the manufacturer of the packaging or the importer of either the packaging or the products; this system, which applies the concept of 'Producer Responsibility', can be found, for example in France, Germany and Sweden. The concept of Producer Responsibility was in fact, incorporated into Germany's 1991 Ordinance on packaging waste because it was viewed as a way of implementing the Polluter-Pays principle (Haverland, 1998). The single-point obligation is considered to be quite effective, as the reduced number of responsible parties reduces the monitoring and enforcement costs. However, the concept of Producer Responsibility has been highly criticised by industry for placing sole liability on the manufacturer/importer because this make it accountable for activities over which it has little control (ERRA, 1998). The single-point obligation is also disturbing with regard to the Directive, which speaks of 'shared responsibility' among the packaging chain members.

Notes

1 Council Directive 85/339/EEC of 27 June 1985 on packaging for liquid foodstuffs, *Official Journal* No. L176, 6 July 1985.

2 Commission *v* France, Case 255-93, 5 October 1994 [1994] ECR 4949.

3 The Commission of the European Communities *v* The Kingdom of Denmark, Case 302/86, 20 September 1988, [1988] ECR, p. 4607. This case is further discussed in Chapter 4.

4 It should be noted in retrospect that the government never viewed point of sale collection as a long-term option: 'the take-back obligation was from the very beginning a stick to force industry to establish its own comprehensive waste management... system' (Haverland, 1998, at p.95).

5 Federal Constitutional Court, Case 1500/93 (1993).

4 The Debate Regarding Constraints on Certain Types of Packaging

Introduction

Soon after the enactment of the Directive in 1994, one prescient commentator noted, 'It seems likely that the Packaging Directive's difficult gestation may have been a harbinger of a troublesome childhood'.[1] Indeed, one of the most contentious issues in packaging waste legislation, both before and after the enactment of the Directive, has been the question of whether a Member State can enact legislation which imposes a higher level of environmental protection, if such legislation interferes with the internal market or distorts competition. Although there is some case law on this point, its value as a precedent and its applicability post-Directive remains unclear. At present, two types of constraints have caused the most controversy: a positive constraint by the German government designed to promote the use of refillable/reusable beverage containers, and a negative constraint by the Danish government in the form of a ban on metal cans and certain types of plastic packaging for beverages.

This chapter begins with an overview of the provisions of the German legislation regarding reuse and analyses it in terms of the Treaty, the Directive itself, and case law. Next, the Danish ban on metal cans is considered in light of Community law. This chapter concludes with a look at the current technological issues and competition policy issues which have contributed to this debate.

Germany's packaging waste legislation

The *Verpackungsverordnung* (the 'German Ordinance') was enacted in 1991; in October of 1996, it became subject to the *Kreislaufwirtschaftgesetz,* a framework law on waste. This framework law established a waste management hierarchy in which prevention has the highest priority, followed by reuse and then recycling. Incineration is permitted, as a last resort, but only if the recovery of energy is its main purpose (ERRA, 1998). As the result of allegations that the German Ordinance interfered with the internal market, the German government has been working with the Commission on revisions to the Ordinance since 1993. In December of 1995, the German Government notified the Commission of its latest revisions. In response, Finland, France, Italy, Spain and the United Kingdom sent Detailed Opinions, objecting to the reuse quotas contained therein (Long et al., 1997).

Also in December of 1995, the Commission sent its first letter of formal notice to Germany, alleging that the German Ordinance contravened EC rules in that it infringed Article 30 of the Treaty (EIS, 1996a). One should note that at this point, there were no objections by the Commission that the Ordinance was in violation of the Directive, only that it was infringing the Treaty. Since that time, discussions between the Commission and the German government have expanded to include the Directive; however, the Commission has not begun any infringement proceedings against Germany for either an infringement of the Treaty or of the Directive.

The Revised Ordinance was rejected in April 1997 by the *Bundesrat*. Following more negotiations within the German government, a new proposal (the 'Novelle') was adopted by both houses in the summer of 1998. Although the Novelle now permits the use of milk pouches for single use, it still contains the quotas on reusable beverage containers and apparently contains sectoral quotas that have not yet been discussed with the Commission. Thus, in November of 1998, the Commission sent Germany a Supplemental Letter of Formal Notice, which requests more detailed information. The Commission thus continues to make it clear that despite its willingness to discuss the situation, final infringement proceedings before the Court of Justice remain a distinct possibility.

Like other Member States, Germany requires manufacturers, distributors and importers of packaging to 'take back' the packaging after use, an obligation which can be met either via individual collection and treatment, or by joining a co-operative scheme. However, with respect to beverage containers, Germany's legislation is unique. Contrary to popular belief, the German Ordinance is actually based on a deposit system, which imposes a DM 0.50 deposit on each one-way beverage container up to 1.5 litres and a 1 DM deposit thereafter. However, firms are exempted from the deposit system where: 1) industry achieves a reuse quota of 72% on refillable containers for liquids such as soft drinks, beer, wine, water and juice (and a 17% reuse quota on containers for milk, which will be raised to 20% by the Novelle); and 2) the firm participates in the *Duales System Deutschland (DSD)* recycling/recovery scheme. Although there was originally a provision

that the reuse quota of each *Länder* must not drop below its 1991 level, this has now been deleted, following discussions with the Commission. Since 1991, the refill rate has wavered just above 72% and thus, the deposit system has not yet been imposed. However, it is important to note that as of 1989, a compulsory deposit has been charged on plastic bottles; these single use containers with deposit are *not* excluded from the 72% quota. As a result, certain mineral water importers (such as Evian and Vittel) have lost a substantial portion of their German market share (Europen, 1998a, 1998d).

Market data in recent years has indicated that nearly half of the increase in one-way packaging has been accounted for by new drinks, which are used for consumption away from the home. However, this increase has led to a decrease in refillable containers in Germany such that the quota has almost not been met (ERRA, 1997a). Industry, which has had extensive problems with the quota system, is even more concerned by the deposit system; it thus wants the quota/deposit to be declared invalid, as the financial repercussions of the implementation of the deposit system would also be substantial, i.e., a deposit of DM 0.50 would double the price of a soft drink (Europen, 1998d).

In early 1997, a compromise to the refillable quota was proposed, whereby specific foreign wines and mineral waters would be exempted; however, the reaction of industry was so negative that this idea was later abandoned. The German government has also been considering alternatives to the controversial system: it has suggested the issuance of tradable permits for non-refillable packaging. Industry has however, rejected this idea with equal force, as tradable permits would have virtually the same impact upon Member State trade as the refill quotas (ERRA, 1998). Since then, another suggestion has been the adoption of the reuse provisions of the Netherlands (which have not been challenged). Under these provisions, existing refillable packaging can only be replaced by one-way packaging if it can be shown that the one-way packaging has the same or a lesser impact on the environment. The existing refillable systems are not however, permitted to decrease by more than two percent. Industry has not reacted strongly to this proposal, likely because recent studies have begun to suggest that one-way packaging may have a lesser impact on the environment.

Relevant European Community law

In analysing the legal implications of packaging legislation, one must bear in mind the general hierarchy of European Community law: the Treaty takes precedence, followed by the provisions of directives and regulations, all of which are interpreted by the case law of the European Court of Justice.

Provisions of the Treaty

While several articles of the Treaty affect the interpretation of the German Ordinance, the analysis of the Ordinance naturally focuses on Articles 30-36, which provide the general parameters of trade among the Member States.

Articles 30-34: the general rules

The provisions of Articles 30, 31 and 34 prohibit quantitative restrictions on imports or exports, or any measures having equivalent effect, and forbid Member States from introducing any new quantitative restrictions or measures which would have an effect equivalent thereto. This rule was interpreted by the European Court of Justice in the well-known 1979 case, *Cassis de Dijon.*[2] The general principle of *Cassis de Dijon* is that a product which is legally manufactured and sold in one Member State must be admitted to the market of any other Member State. The rule was further expanded in the 1982 case, Commission *v.* Ireland,[3] where the ECJ held that either an interdiction or a incentive can be considered as a discriminatory measure (in that particular case, a campaign in favour of national products).

It seems clear that the German Ordinance is an infringement of Article 30: the quota in favour of reusable beverage containers results in a reduced number of imported beverages and restricts the amount of one-way containers to 28% of the market; accordingly, this has an effect equivalent to a quantitative restriction. Indeed, the German Ordinance has been described as constituting 'a *prima facie* restriction of the free movement of goods' within the Community (Long et al., 1997). And in fact, Directorate General XV of the Commission, in its 6 September 1995 communication to the Austrian authorities, indicated that a statutory preference for reusable packaging should be considered as a measure having equivalent effect to a quantitative import restriction, as defined by Article 30 and thus, is not in conformity with the Directive (Commission Hearing, 1997).

With regard to the *Cassis de Dijon* ruling, it can be observed that one-way packaging which is legally for sale (as part of a product) in other Member States must be admitted to the market of Germany, another Member State. Per the ruling of Commission *v.* Ireland, the fact that the German quota is in effect, an incentive (to promote reusable packaging) does not render it non-discriminatory. Indeed, in arguments submitted in the *Danish Bottles* case *infra*, the Commission argued that reuse systems favour operators which are close to the market and are *de facto* discriminatory, or at a minimum, disproportionate. Thus, an analysis of Articles 30-34 by themselves indicates that the German quota infringes these articles of the Treaty. However, one must next consider the exceptions permitted thereto.

Article 36: the exceptions

According to the foregoing, the general rule with respect to environmental legislation seems to be that a Member State cannot introduce legislation more strict than that of the Community, whether to protect the environment in its own territory, or that of another Member State. In this instance, it would be an infringement of the Treaty were a Member State to impose an import restriction on a certain type of material deemed harmful to the environment, or an export restriction on a product whose sale is prohibited in that Member State, but not in others (even where such a restriction would protect the environment). However, other articles of the Treaty and the case law of the European Court of Justice would suggest otherwise.

The restrictions of Articles 30-34 are subject to the exceptions of Article 36, which allow such prohibitions or restrictions where they are justified on the grounds of, *inter alia,* the protection of the health and life of humans, animals or plants. Although the Court of Justice has historically interpreted this clause quite narrowly and has held that the list of exceptions is inclusive, the general protection of the environment has been considered a sufficient basis. Yet, Article 36 tempers these exceptions by adding that these exempted prohibitions or restrictions cannot serve as a method of arbitrary discrimination or a disguised restriction on trade between the Member States.

In its December 1995 infringement letter to the German government, the Commission stated that the German Ordinance gave rise to trade barriers and thus infringed Articles 30-36 of the Treaty. The Commission indicated that the trade barriers were created by the Ordinance's preference for reuse via refill quotas and the costs of recovery associated therewith. In the German government's response of April 1996, it replied that the Ordinance did not infringe the Treaty because the legislation was applied equally to national and foreign producers. The government admitted however, that 'a negative impact on intra-Community trade cannot be totally excluded'.[4] The German government stated that the provisions were nonetheless justified as measures necessary for the protection of the environment: eco-balances had established the advantages of refillable packaging over one-way packaging as a measure contributing to the prevention of packaging waste. Further, the prevention of packaging waste was an important basis of the German Ordinance, due to problems of increasingly limited landfill space in the Federal Republic. Without such refillable quotas, there would be a shift from refillable packaging to non-refillable packaging, leading to an increase of waste generated by beverage packaging. Thus, Germany concluded, both the 72% refill quota and the underlying deposit system were justifiable, pursuant to Article 36 of the Treaty.

However, the arguments of the German government are misplaced for two reasons. First, while the original rationale for the quotas was that Germany's available landfill space was rapidly decreasing, since that time, the estimates of available space have been revised; today, even though the total amount of drinks packaging has increased, the amount of landfill used has decreased over the same period and today, there is an alleged shortage of waste for some treatment infrastructures (Europen, 1998d). In any event, this argument was always questionable, as the refill quota only applied to one type of one-way packaging, that of some beverages. Further, the original arguments no longer have merit as the recycling targets of the original Ordinance have been fulfilled and industry has succeeded in reducing the amount of packaging waste through source reduction (ERRA, 1997a). Thus, the environmental justification for the German Ordinance is no longer valid. Second, the arguments of the German government fail to take into account the provision of Article 36 that even when the measure is justified on environmental grounds, it cannot be a method of arbitrary discrimination or a disguised restriction on trade.

Method of arbitrary discrimination? Despite Germany's contention that the refill quota is applied equally to foreign and domestic firms, it is clear that a requirement that a firm take back and treat its containers so that they can be refilled discriminates against foreign producers, which must transport the containers for greater distances. Statistical evidence collected since 1991 has clearly shown that as the reuse percentage drops toward 72%, the percentage of imports correspondingly decreases, in order to guarantee that the deposit system is not implemented. Similarly, in years when the reuse percentage has risen above 72%, the pressure on imports has decreased, and their numbers have risen (Europen, 1998d).

In another example, there remains the issue of approved containers for certain beverages: while German wine producers allow for 250 standard wine bottles which can be refilled, France uses over 750 different bottles; Spain, Italy and Portugal are in a similar situation (Commission Hearing, 1997). Moreover, it makes no difference that there is an alternative of a deposit system: such a system would be primarily used by foreign producers and the reluctance of consumers to pay a deposit would subsequently result in a smaller market share, such that placing a drink on the market would eventually prove economically impracticable. Thus, the German Ordinance qualifies as an arbitrary method of discrimination.

Disguised restriction on trade? While the Community institutions are told that the protection of the environment is the justification for the quota, within Germany another story is emerging. Several trade associations and politicians have openly voiced support for the quota, referring to it as 'SME friendly' and praising it as a good barrier to imports which would threaten jobs in the brewing industry. The German Ordinance has also been lauded for protecting both regional and local enterprises. Such public opinions often approvingly note that the mandatory deposit system would make it difficult for foreign breweries to export their beer to Germany in metal cans (Europen, 1998d). It is thus evident that whatever the officially stated environmental motivations behind the German Ordinance, there is an additional protectionist element at work, reflecting a clear violation of Article 30 of the Treaty. Accordingly, upon a strict analysis of the wording of Article 36, the German Ordinance does not fall within one of its exception.

Protection of only the German environment? It is worth noting at this point that although the German government has nobly cited the protection of the environment as its justification for preferring reusable containers, this protection appears to extend only to the German environment. For, while 1996 figures for the German beverage market indicate that more than 70% of the German beverages sold in Germany are sold in reusable containers, 94% of German beverages sold outside Germany are sold *in one-way packaging* (Europen, 1998d). Thus, the special concern for protecting the environment exhibited by German officials appears to only apply to the German landscape. Moreover, it is not a coincidence that the German government does not impose its rigorous quota on German beverages exported, as it knows that the logistics of establishing a system for treatment of reusable packaging outside the country of product origin are very difficult.

Articles 130r-t

These articles, which were significantly modified by the Maastricht Treaty, have served as the basis of an argument that Member States are permitted to enact a higher level of protection, even where that protection may interfere with Member State trade. For example, Article 130t states that Member States shall not be prevented 'from maintaining or introducing more stringent protective measures'. The theory is that because the Directive does not specify a precise system for reuse, in that it establishes neither quantitative objectives for reuse, nor methods for achieving them, Member States are allowed to establish whatever type of system they wish, even if it is a system which interferes with Member State trade (Demey et al., 1996). Yet, one must note that the Maastricht Treaty added the following two provisos to Article 130t: 'Such measures must be compatible with this Treaty. They shall be notified to the Commission'. Thus, while it *is* correct to state that Member States have discretionary power with regard to establishing more strict environmental protection objectives, the same discretionary power does not permit Member States to pursue any type of reuse system they wish (despite the lack of more precise specifications in the Directive), where such a reuse system is not compatible with the Treaty.

Articles 12 and 95

These two articles also affect the free flow of goods in the Community. Article 12 et seq. prohibit the imposition of any kind of customs duties or any charges having equivalent effect. Here, it can be argued that the deposit scheme underlying the quota is a charge having equivalent effect, as it is more likely that foreign producers, which have more difficulties meeting the quota, will become subject to the deposit scheme. As a result, the quota and deposit scheme impose greater charges on foreign producers that in effect, are a tax on a product which is in excess of the tax imposed on similar domestic products, as prohibited by Article 95 et seq. Thus, the German Ordinance infringes the Treaty with respect to these articles.

Articles of the Directive

Several provisions of the Directive address the issue of reuse and of trade among the Members States. While Article 5 has the greatest direct impact, Articles 9 and 18, as well as the Preamble, are also noteworthy. However, one must first consider the issue of prevention, as it relates to reuse.

Priority of prevention

As Article 1(2) makes clear, the prevention of packaging waste is of primary importance to the Directive, in keeping with the priority of prevention in the European hierarchy for waste treatment. The Commission has long considered that a system based on a preference for reuse contributes to a reduction of packaging produced and thus, is in conformity with the Directive. (One should note that in contrast, other methods of waste management, such as recycling and energy recovery, only reduce the amount of packaging waste which would otherwise be landfilled). Yet although a system of reuse preference would be in compliance with the Directive, the Commission has also acknowledged that it must be in conformance with Community law, particularly as regards barriers to trade and distortions of competition (Commission Communications, 1997). Further, it is important to note that nowhere in the Directive is 'reuse' classified as a method of prevention or placed on a higher priority than recycling or recovery.

Article 5 of the Directive

This Article specifically addresses reuse:

> Member States may encourage reuse systems of packaging, which can be reused in an environmentally sound manner, in conformity with the Treaty.

A review of Article 5 thus indicates that while Member States are permitted to implement systems to encourage reuse, 1) said packaging must be suitable for reuse in an environmentally sound manner; and 2) the reuse system must not infringe the Treaty.

Packaging suitable for reuse in an environmentally sound manner At first glance, it would appear that one-way packaging would be more detrimental to the environment than reusable packaging because one-way packaging means just that: the packaging travels from manufacturer to consumer and is then discarded and must be treated as waste. With certain materials, this treatment usually involves landfill or incineration. In contrast, reusable packaging can be refilled several times before it has to be treated as waste.

It was on the foregoing assumptions that the German Ordinance and the Directive were based. However, research conducted during the past few years suggests that with respect to reusable containers for liquids, reusable packaging may only be more beneficial if certain conditions are met; under other circumstances, it has no greater environmental benefit than one-way packaging and indeed, can even be more harmful to the environment.

These assumptions were partly caused by the fact that the early German studies were based solely on waste generation; they did not assess the total impact on the environment of factors such as increased energy consumption and effluents into water and air as the result of the transport and treatment of the reusable packaging. In some regions, the toll of reusable packaging on the environment is greater than if one-way packaging had been used and then treated as waste. For example, in northern Finland, recycling efforts have been abandoned, as the cost (financially and to the environment) of transporting containers from rural areas to treatment facilities near populous Helsinki outweigh the potential benefits. Further, the data from the early German studies, which only concerned beer and milk, was then extrapolated and applied to all types of beverage containers. Thus, what constitutes an 'environmentally sound manner' remains open to debate.

Reuse systems of packaging in conformity with the Treaty Although this second condition of Article 5 seems rather straightforward, it too has engendered controversy. While it might at first glance seem that the German quota would not conform to the Treaty in that it creates barriers to trade, it may also be argued that as the Treaty permits each Member State, in the absence of Community provisions to establish its own environmental policy objectives, a higher level of environmental protection is permissible (Demey et al. 1996). But the Commission itself, in its Detailed Opinion regarding Luxembourg's proposed legislation, noted that systems which prefer reuse are more likely to have a greater negative effect on imported than on domestic production and thus, could be in violation of the Treaty (ERRA, 1997a). It appears therefore that the decision regarding reuse systems' conformity with the Treaty must be made on a case-by-case basis, according to the particular circumstances. Accordingly, the decision ultimately comes down to the classic balancing of the protection of the environment against the correct functioning of the internal market.

Preamble of the Directive Several recitals in the Preamble relate to the issue of reuse. Recital 7 states that 'the management of packaging and packaging waste should include as a first priority, prevention of packaging waste and, as additional fundamental principles, reuse of packaging, recycling and other forms of recovering packaging waste and, hence, reduction of the final disposal of such waste'. Thus, reuse is identified as a fundamental principle, but on the same level as recycling and recovery. Recital 10 notes that Member States may encourage reuse systems for packaging, when the packaging can be reused in an environmentally sound manner.

However, the foregoing recitals are tempered by two others. Recital 8 provides that until scientific progress indicates otherwise, reuse and recycling should both be considered preferable with regard to their impact on the environment; Recital 8 also notes that life-cycle assessments should be completed as soon as possible, so that a hierarchy can be justified. Recital 15 states that, using the reports of the Commission, both the Parliament and the Council should 'examine the practical experience gained in Member States in working toward the aforementioned targets and the finding of scientific research and evaluation techniques such as eco-balances'. Thus, the Community institutions are expected to take into account the research conducted post-enactment regarding the best methods of waste treatment.

Alleged need for restrictions to promote reuse One should note that the use of the phrase, 'encourage reuse systems of packaging', does not necessarily mean the use of quotas. Indeed, there are other methods of encouragement, such as: information campaigns to educate both industry and the consumer; regulations requiring retailers to carry products in both one-way and reusable packaging; the promotion of environmental agreements among members of industry and government to encourage reuse; regulations forbidding industry from having different prices for the same product, based on one-way versus returnable packaging; and fiscal incentives (Europen, 1998d). Yet, Germany has failed to try any of these methods.

Articles 9 and 18 of the Directive

Article 9, which specifies the procedure for determining the Essential Requirements of the Directive, states that a Member State cannot discriminate against the importation of packaging which is in compliance with said Essential Requirements. Article 18 provides that a Member State cannot impede the introduction of packaging into its territory if the packaging complies with the Directive. Given that one-way packaging does meet the Essential Requirements of the Directive and is in compliance with it, both of these articles strongly indicate that it is not permissible for Germany to ban such packaging. Germany has argued that it does not ban one-way packaging, but instead indicates a preference for reusable packaging. Yet, in light of the fact that preferences for reusable packaging intrinsically favour locally manufactured packaging, and the fact that retail chains in Germany have begun to delist beverages in one-way packaging (Europen, 1998a), the German quota system may be considered a *de facto* ban on such packaging, and thus forbidden by the Directive.

Articles 2 and 23 of the Directive

Article 2 states that the Directive is applicable to all types of packaging, while Article 23 specifically repeals Directive 85/339/EEC on containers of liquids for human consumption, such that the only European legislation affecting liquid drinks containers is the Directive. Yet, one interesting aspect of the German quota is that it is only targeted toward beverages. To date, there has been no justification as to why milk receives a special quota of only 17%; however, it is clear that as fresh milk is a domestic product, this lower quota puts less pressure on domestic producers. Further, it is evident that there is no justification for the sectoral nature of the quota: if one-way packaging is arguably bad for the environment, it should be bad regardless of its contents. Moreover, only glass presently complies across the board with the requirements of the Ordinance; PET bottles can only be used for soft drinks. Thus, the quota is in violation of the Directive to target one specific type of container or type of material, and indeed, creates barriers to trade by so doing.

Does the Directive give Member States greater freedom regarding reuse than with other methods of waste treatment?

It has been suggested that because the Directive fails to expressly stipulate that Member States' systems of reuse must not cause distortions to the internal market or create technical barriers, Member States have been given significant latitude in the creation of such a system. It has been further suggested that this latitude, when combined with the principles of prevention and reduction of source as outlined in Article 130r, make systems of reuse less susceptible to criticism regarding their effects on the free flow of goods (de Sadeleer 1995b). Yet, such an argument fails to acknowledge the fact that all measures in EC law are always equally subject to the fundamental freedoms of the Treaty, in this case, the free movement of goods. It is only when a system of reuse begins to affect the free flow of goods that one can begin to look for exceptions and exemptions which are based on the protection of the environment and yet subject to certain restrictions, as discussed *supra*.

The idea that because the legislator did not specifically state in Article 5 of the Directive that systems of reuse must not cause distortions to the internal market, and that therefore, such systems are less subject to one of the fundamental rules of Community law, is an argument of semantics. Indeed, the legislator *did* state in Article 5 that systems of reuse must be 'in compliance with the Treaty'. And it is clear that to be in compliance, the legislation must comply with the fundamental freedom of the free movement of goods, subject of course, to those exemptions discussed *supra*. Thus, systems of reuse *are* subject to the rules of the internal market, just as is any other type of legislation enacted in the Community.

Moreover, one must acknowledge that the German Ordinance is, in effect, in contradiction with the purpose of the Directive, the idea of which was to harmonise the disparate packaging waste legislation of the Member States while maintaining a high level of environmental protection. Although it is clear that environmental protection is an important aspect of the Directive, it cannot be pursued to the exclusion of the equally important aspect, namely, the smooth functioning of the internal market.

Applicable Case Law

The only case of the European Court of Justice which specifically addresses packaging is the Commission *v* The Kingdom of Denmark,[5] better known as the *Danish Bottles* case. The original Danish legislation which led to the decision was a 1978 law that created a deposit system for the return of containers for beer and soft drinks, which could only be sold in reusable packaging. All containers had to be approved by the Danish Environmental Protection Agency, which could withhold approval where an appropriate package of the same volume had already been approved.

As the requirement for packaging approval began to affect importers, complaints were made to the European Commission, which began discussions with Denmark. In March of 1984, Denmark enacted Decree 95, which allowed the use of non-approved packaging (with the exception of metal packaging) as long as: 1) a deposit system had been set up for this packaging; and 2) a yearly limit of 3,000 hl per producer was not exceeded.

Thereafter, industry complained that even the revised law was a barrier to trade because of the cost and logistical problems. The Commission thus filed an infringement proceeding, which was ultimately heard before the Court of Justice. The Court upheld the part of the Danish legislation regarding the creation of a deposit system, ruling that barriers to the free movement of goods within the Community which are due to differences in national laws *are* acceptable: a) 'in the absence of common rules relating to the marketing of the products in question' and b) insofar as the barriers: 1) are applied indiscriminately to domestic and imported goods; 2) are necessary to achieve a fundamental requirement of Community law (such as the protection of the environment); and 3) are proportionate to the goal sought, in that the barriers constitute the minimum restriction possible to the free movement of goods.

Absence of common rules Although the *Danish Bottles* case has been cited as support for Germany's reuse system of quotas, in fact, it is not clear that the case has such value. One must look to the initial conclusive statement of the Court's ruling: '...in the absence of common rules relating to the marketing of the products in question...'. At the time of the *Danish Bottles* decision, discussions regarding a possible directive on packaging waste had not even started and the return and recovery schemes of today had not yet even begun. Clearly, one can argue that the Directive itself is a 'common rule' which must be respected and that this common rule states that the purpose of the Directive is to not only provide a high level of environmental protection, but also to harmonise to the extent possible, the heterogeneous packaging legislation which then existed in the Member States. If one accepts that the Directive is a common rule, then one must look strictly to the Directive and the Treaty and cannot apply the ruling of *Danish Bottles,* as the Court has stated that the ruling is only applicable in the *absence* of common rules.

Impartial application On the other hand, one can also argue that because the Directive does not specifically regulate the marketing of reusable packaging (other than to state that the reuse systems must be in conformity with Treaty), then the *Danish Bottles* ruling would apply. The first requirement thereof is that the measure be applied impartially to domestic and foreign producers; in this respect, Germany is correct in that the German Ordinance is applied equally to domestic and foreign producers. Although there has been some argument made that the law is applied unfairly to foreign producers, this argument is misplaced: the *effect* of the law may be unfair, but the law itself has been applied equally to all manufacturers of containers for beverages.

Necessary barrier? Until recently, it would have been presumed that the German Ordinance easily met the second requirement of *Danish Bottles,* i.e., that the reuse quota was necessary to protect the environment. But industry has begun to contest whether the quota is even necessary to protect the environment, given that one of its primary rationales, to prevent packaging waste so as to reduce the amount of landfilling required, is no longer a serious problem in Germany (EIS, 1997a). Moreover, as discussed *supra,* only the German environment is allegedly being protected by the measure.

Proportionality Yet, even if one accepts that the German Ordinance is necessary to protect the environment, the Ordinance must still pass the test of being proportionate to the goal sought, such that the Ordinance constitutes the minimum restriction to the free movement of goods. To date, Germany continues to insist that the quota/deposit system is the minimal system possible to ensure reuse of beverage containers and that the withdrawal of the requirement would result in a significant decrease in the use of refillable containers. Yet, as of 1995, only 12.63% of the drinks imported into Germany were in refillable containers, whereas 75.09% of domestic drinks were in such containers (ERRA, 1997a). Moreover, the figures of *DSD* for 1996 indicated that the recycling rate had at that time already exceeded 80%, thus raising the issue of the need for further reuse quotas. Interestingly, *Gesellschaft für Verpackungsmarktforschung* ('GVM'), a German research institute, completed a study in the spring of 1997 which concluded that even if the refillable quota was eliminated, the market share of refilled bottles would still be 67.4% by the year 2000. One should note however, that the German government has contested this study as statistically flawed. While Directorate General XI of the Commission appears to concur with this opinion, other Directorates are not in agreement (Commission Hearing, 1997; Europen, 1998a, 1998c).

Ironically, the most simple alternative would be to exempt all imported beverages from the reuse quota and deposit system. Given that the market share of imported beverages is less than 10%, the effect would be minimal (Europen, 1998d). Of course, one must acknowledge that following such an exemption, imports would naturally increase and thus, have a greater effect.

But the German government could certainly conduct a trial of the exemption (for example, for a two year period) and then analyse the results.

The claim that the German reuse system is the minimum restraint possible was further diminished by the conclusion of the above-referenced GVM study that the result achieved by *DSD* of a reduction in packaging consumed (due to developments in packaging, such as the introduction of lighter materials) was much more significant for the overall goal of preventing packaging waste, than any of the effects caused by the quota on reusable containers (Europen, 1998a). This study thus suggests that leaving the reduction of waste to a national recovery/recycling system is a viable alternative for the minimal restriction test of *Danish Bottles.*

Alternatives in other Member States While Germany may insist that the reuse quota is the minimum restraint possible to promote reuse, other Member States have found alternatives. Austria, for example, exempts reusable packaging from some of its waste reporting requirements and uses a system of combined targets to permit the market for waste to determine the most economically efficient choices from among refilling, recycling and energy recovery. Sweden uses similar combined targets and although it does have a 95% reuse target, it does not apply to imported containers (Europen, 1998d). Given the range of successful options used by other Member States and Germany's failure to try any of them, one begins to suspect that protectionist motivations play a greater role than environment ones with regard to the continued imposition of the reuse quotas.

What will happen if the German reuse system is approved?

Should the Commission approve the German quotas or fail to take any action at all, it is possible that other Member States will follow suit, if for nothing else than to protect their national markets to the extent permitted by the Treaty, as per the Commission. Yet, even if the Commission permits the German quota, there remains a range of options to challenge the decision. For example, it is likely that another Member State would bring an action against Germany on the grounds of Article 170 of the Treaty, which permits a Member State to bring an action against another Member State for failure to fulfil an obligation under the Treaty. Further, an individual or entity could challenge the German quota locally and asked the national court to refer the matter to the ECJ under Article 177 (Long et al., 1997).

Implications for the Danish ban on metal cans

Although Denmark continues to impose its ban on one-way beverage containers and cans, several Member States have complained and the Commission has launched an investigation. In 1996, Denmark adopted three decrees regarding packaging, which were then notified to the Commission. In addition to the Commission's protest that the decrees should have been notified before their adoption, several Member States filed Detailed Opinions. On November of 1995, multiple complaints were filed with the Commission, challenging both the ban on one-way beverage containers and the ban on aluminium and steel cans. As of May 1998, Denmark had removed the ban on aluminium beverage cans, as long as a recycling rate of 90% was attained. However, Denmark continues to maintain a ban on steel beverage cans and one-way beverage containers. Moreover, these bans are an infringement of the Directive because they ban packaging which meets its essential requirements (ERRA, 1998).

Although Denmark may argue that the *Danish Bottles* decision supports its ban on metal cans, this assertion is invalid for two reasons. First, *Danish Bottles* requires that there exist no 'common rules' regarding the marketing of the product. It is true that at the time of the *Danish Bottles* decision, there was no Community law on the matter; however, the Directive is now applicable and it does not permit discrimination on the basis of packaging materials which are used in other Member States. Second, even if one accepts that the *Danish Bottles* ruling still applies to Denmark, Denmark must continue to demonstrate that the measure is: 1) necessary to product the environment (a questionable claim, given that more than 80% of Danish household waste is incinerated and all incinerators permit magnetic extraction (EIS, 1997a)); and 2) proportionate to the goal sought, in that it is the minimum restriction possible (again, a doubtful assertion, in light of the other methods available to promote reuse). Thus, although in terms of general Community law, the *Danish Bottles* rules remain valid, the ability of Denmark to meet them is questionable. The Commission has brought an infringement proceeding against Denmark, but is continuing discussions with its government. The Community's decision regarding whether to pursue an infringement action against Germany will likely play a part in this decision.

A question of technology

Before discussing the precise technological issues, it is useful to distinguish among the three types of reusable packaging systems being used in the Community:

1) reusable transport packaging, i.e., between the manufacturer/ distributor and the retailer;

2) home refill, i.e., the product is sold in light, simple packaging which is designed to be taken home and transferred to a permanent container or dispenser; and

3) return/refill store systems, i.e., either: a) the container is returned to the store, from where it is transported for treatment and then is refilled; or b) the container or utensils are used on the premises, where they are then treated and reused (for example, in fast-food restaurants).

It is the third type of packaging that has been subject to the most extensive legislation in the Community thus far (Europen, 1997).

Technological advances - the debate regarding plastics versus glass

Articles 19 and 21 assist with the adaptation of the Directive to changes in technology; it is clear that these articles will be utilised more in the future as research and product development continue to rebut classical assumptions about packaging. One of those traditional assumptions is that glass is always more environmentally-friendly than plastics. The Directive itself is based on this assumption, with the tables of Annex I listing the quantities of waste collected in tons, as opposed to volume; this preference toward weight favours glass. Moreover, reuse systems favour glass, as it can be reused more often than plastic; while the newer PET plastic bottles can be reused fairly often, they still have a more limited life-span than glass (ERRA, 1997a).

Yet, recent studies have begun to challenge the assumption of superiority of glass over plastics. While it is true that glass can be reused whereas plastics are thrown away, one must consider the costs of reuse, as well as the fact that eventually, the glass container must also be disposed of. Furthermore, a 1995 study by the consultancy Ecobilan SA, concluded that the German government's reliance on the amount of waste generated as the justification for the Ordinance was misplaced, given that other factors involved in transporting the containers, such as air and water effluents, and fuel usage, could have an even greater impact on the environment.

The use of distances to determine exceptions

In response to the controversy surrounding reusable containers, one solution proposed by the German government was to create break-even transport distances, within which a container must be reused. Yet, a study of the Frauenhofer Institute demonstrated that there can be substantially different effects on the environment, using break-even distances, because the parameters vary so dramatically. Thus, one obtains different results between fresh milk in Austria and that in Germany; between fruit juices and fresh milk; and even between fresh and UHT milk. The study thus demonstrated that break-even transport distances are rather subjective, due to differences in factors such as the type of product and its intended use, its packaging, etc. (UNEDA, 1997). Accordingly, industry protested vigorously and this idea was dropped. However, it has been acknowledged that reusable packaging generates less waste than one-way packaging in those situations where a high trippage rate is achieved and where the container only travels short distances for treatment (Europen, 1997). Thus it would, for example, be environmentally sound for Germany to require its *domestic* producers to utilise reusable beverage containers, subject to certain break-even distances.

Life-cycle assessments and Eco-balances

Life-cycle assessments and Eco-balances in general The Preamble of the Directive states that 'life-cycle assessments should be completed as soon as possible, to justify a clear hierarchy between reusable, recyclable and recoverable packaging'. In addition, Article 10 commits the Commission to 'the preparation of European standards relating to the criteria and methodologies for life-cycle analysis of packaging'. Life-cycle assessments (LCAs) are popular because they further the goal of sustainable development to reduce the global environmental burden caused by the provision to society of goods and services. LCAs have thus been used to focus on those areas where significant improvements can be made, by using a 'cradle-to-grave' approach which looks at the overall impact of the good or service on the environment. LCAs present two advantages: first, as LCAs use a cradle-to-grave approach, they help to avoid 'problem shifting', i.e., the situation in which an improvement in one aspect leads to an increased environmental burden in another. Second, LCAs attempt to develop a ratio between the value of a product/service and its impact on the environment.

There are however, several limitations to LCAs which must be acknowledged. Among them are the fact that LCAs can rarely demonstrate a clear-cut choice among various systems; instead they provide an input of potential benefits and costs, depending on the local situation. In addition, an LCA only relates to one product/service at one point in time; its use in calculating future situations is limited. Moreover, an LCA does not of course, consider non-environmental factors, such as economic or social issues, all of which should be taken into account when a political decision is made. Lastly, one must bear in mind, that the standards and methodology of LCAs remain under development. That having been said, LCAs are being used more frequently as a supplementary information tool, including in the packaging waste sector (Europen, 1996).

Similarly, Eco-balances have limited advantages and disadvantages and should not be the sole, or even the primary, basis of a political decision. For while it is true that Eco-balances facilitate decision-making, they are dependent upon various factors such as local, regional or national usage, the material and the relevant market, all of which may change rapidly. Thus, for example, while Eco-balances may reveal a preference for reuse versus recycling with regard to glass, or that for aluminium or steel recycling is preferable to incineration with energy recovery, Eco-balances do not demonstrate preferences with respect to materials such as paper and fibreboard (Eco-balances, 1997).

Life-cycle assessments and Eco-balances in support of policies Germany has used both LCAs and Eco-balances to support its contention that reusable packaging is better for the environment than one-way packaging. Yet, given the variations of packaging, from product use to regional differences, it is extremely difficult to rely on a few LCAs. One trade association has noted that of approximately 20 LCAs conducted during the last decade, 15 have indicated that there is no preference for one system over the other (Commission Hearing, 1997). Ironically, both the original and subsequent LCAs commissioned by the German environment agency stated that one cannot allege the superiority of one type of system over another using only LCAs (Europen, 1998d).

Indeed, a Dutch study concluded that even in the Netherlands, a country much smaller than Germany, it was impossible to identify one method of packaging treatment as superior to another, and that such an analysis can only be conducted on a product by product basis (Europen, 1996). Moreover, a 1997 study on Eco-balances for the Commission concluded that while one-way beverage cartons and PET bottles seem to have the least impact on the environment for both small and large liquid packaging, the variations are not significant and that using one or the other would not make a substantial difference (Eco-balances, 1997). Thus, LCAs and Eco-balances alone cannot be used as an environmental justification for reuse quotas.

Competition issues

Another area affected by constraints on certain types of packaging is that of competition policy. As a reuse quota makes it necessary for an importer to arrange for the take-back and treatment of its packaging, this difficulty can become insurmountable for the small importer. The requirement also acts as a barrier to new entrants to the market or those undertakings trying to launch a new product in the marketplace; it bears noting that new beverages are usually launched in one-way containers as the risk of failure associated with any new product makes it impractical to invest in a reusable container treatment system (Europen, 1998d). As a result, the range of products available to the German consumer is unfairly limited, in contravention to the general competition policy of the Community. These problems demonstrate that the German reuse quota is contrary to the goals of the Treaty; of the creation of a single market; and of European Community law in general.

Conclusions

It is clear that the Treaty and European case law permit the enactment of Member States legislation imposing a higher level of environmental protection, even if such legislation results in barriers to trade or distortions of competition. However, the Member State must demonstrate that the measure is not an arbitration discrimination or a disguised restriction on trade, that the measure is necessary and proportionate to the objective, and that no less restrictive measures are available. The German Ordinance has failed to meet these tests. Where the German government could have once successfully argued that scientific evidence justified the quota, more recent scientific research has revealed the fallacy of the assumptions upon which the German Ordinance was based. Seven years of experience with the effects of the German Ordinance as well as developments in packaging research which have led to a reduction in waste, have also rendered meaningless many of the original justifications for the Ordinance.

Similarly, the environmental justifications on which the Danish measures were based are no longer valid. While the European Court of Justice may have rendered a judgment in favour of Denmark and while it is true that the rules of *Danish Bottles* remain valid, Denmark must continue to demonstrate that its legislation meets those rules, a difficult assertion in light of the more recent scientific evidence. The proceedings to come, against Germany and/or Denmark, should establish new precedents.

As the packaging waste landscape has changed during the 1990s, the German Ordinance and the Danish ban have remained static, imposing protectionist measures based on assumptions which are no longer valid. Furthermore, one need only look to one of the primary purposes of the Directive, to avoid barriers to trade caused by the heterogeneous legislation of the Member States, to see that the German and Danish measures are in direct contravention to the rationale for the Directive.

Yet, it remains uncertain as to whether either of these mattes will be resolved by the Commission. Following its enactment in 1994, the contentious nature of the Directive has led to a split of opinion on many of its issues within the Commission itself, with DG III (Industry) and DG XV (Internal Market) often on one side, and DG XI (Environment) and the Commission's Legal Services on the other (EIS 1996d).

Notes

1 Reid, 1995, p. 242.

2 (Rewe-Zentral AG *v* Bundesmonopolverwaltung fur Branntwein), Case 120/78 of 20 February 1979, [1979] ECR 649 [1979] 3 CMLR 494.

3 Case 249/81 of 24 November 1982.

4 Response of Federal Republic of Germany to the letter of Formal Notice from the Commission 29 April 1996.

5 Case 302/86, 20 September 1988, [1988] ECR, p. 4607.

5 The Use of Economic Instruments and Environmental Agreements to Implement the Directive

Introduction

Two interesting methods of implementing the Directive that are expressly authorised in its text are the use of economic instruments and the use of voluntary agreements (now more commonly called 'environmental agreements'). The use of these instruments became more popular during the drafting of the Directive and thus, several Member States pushed for their specific inclusion. This chapter analyses first economic instruments and then environmental agreements, both in a general sense and in terms of how they are being utilised by various Member States.

Economic instruments

References to economic instruments in the Directive

The first reference to economic instruments is in the Preamble, which states that 'it may be appropriate for [both] the Community and the Member States to use economic instruments'. The second reference is contained in Article 15, which is devoted to economic instruments:

> Acting on the basis of the relevant provisions of the Treaty, the Council adopts economic instruments to promote the implementation of the objectives set by this Directive. In the absence of such measures, the Member States may, in accordance with the principles governing Community environmental policy, *inter alia,* the polluter-pays principle, and the obligations arising out of the Treaty, adopt measures to implement those objectives.

In contrast to the Preamble, Article 15 indicates that *the Council* is empowered to adopt economic instruments and then adds that Member States can adopt economic instruments in the absence of such measures at Community level. Thus, although the Preamble implies that both the Council and Member States may adopt economic instruments, the predominant Article 15 makes it clear that Member States can only act in the absence of action on the part of the Council. It is worth noting that to date, all activity has been at Member State level, although Community-wide economic instruments have been discussed for several years. This lack of activity at Community level is likely due to the fact that fiscal instruments must be unanimously adopted.

Article 15 additionally emphasises that all economic instruments must respect the principles of Community environmental policy (for example, the polluter-pays principle) as well as the provisions of the Treaty, while the Preamble notes that the latter must be respected 'so as to avoid new forms of protectionism'. Accordingly, while a Member States is given free rein to impose those economic instruments it wishes, it must nonetheless respect these provisions and principles.

The reference in Article 15 to 'the principles governing Community environmental policy, *inter alia,* the polluter-pays principle' is interesting because it leaves open the use of economic instruments which implement other types of Community policies, such as the 'pollution prevention principle'. During the negotiations of the Directive this article became pivotal, with Belgium, supported by Luxembourg, Germany, the Netherlands and Denmark, refusing to vote for the Directive until Article 15 was revised. The prior version of Article 15 stated that economic instruments 'adopted in accordance with the "polluter-pays" principle, must not lead to distortions of competition, nor obstruct the free movement of goods, nor discriminate against imported products'. Despite assurances of compliance by the Commission's legal service, Belgium insisted that its eco-taxes programme could be challenged, notably as it was based on the idea of 'pollution prevention' and not on the 'polluter-pays principle' (EIS, 1994d). Following the support of other Member States, the abbreviated version of Article 15 was adopted. However, with the exception of the vagueness added by the use of the term *inter alia,* the words deleted from Article 15 are basically a restatement of general principles of Community law and thus, all economic instruments would be bound to respect them.

More interesting was the notification provision of the Amended Proposal.[1] This proposal had added a subsection that would have required Member States to notify the Commission of all economic instruments implemented for purposes of complying with the Directive. In the event the economic instrument involved the distribution of state aids, the Member State would also have been required to notify the Commission in advance of distribution, pursuant to Article 93 of the Treaty. Yet, although these provisions were deleted, the Member States of course remain subject to the provisions of Article 92 et seq. and indeed, the Commission has confirmed that Member States must notify all economic instruments to it (Commission, 1997). Thus in this regard, the Community institutions retain some control over the Member States.

Should future action be taken on economic instruments at the Community level, the issue of whether Member State economic instruments can be continued will depend upon both the type and scope of the Community measures, as well as their legal basis (Demey et al., 1996). Regardless, given the proliferation in recent years of different economic instruments regarding packaging, it will become increasingly difficult for the Community to unanimously adopt an economic instrument in this area (Reid, 1995).

Economic instruments in general

Although eco-taxes and deposit systems have been the primary economic instruments utilised by the Member States to implement the Directive, and thus will be the primary focus of this section, is important to note that there are in fact, several different types of economic instruments that the Member States have employed with respect to products and packaging. Moreover, Member States may have different terms for the same instrument and vice-versa. The following is a description of some of the most commonly used instruments.

Types of economic instruments

Eco-taxes/product fees These are fees charged by the state, often without being earmarked for a specific environmental amelioration. These fees are often designed to discriminate against a particular type of packaging considered harmful to the environment, and/or to push the consumer toward reusable or recyclable packaging. An example of an eco-tax would be the Belgian tax on beverage containers made of PVC.

Mandatory deposits These are systems by which the consumer must pay an additional deposit (often incorporated into the cost of the product) which is reimbursed to him upon return of the container. Although mandatory deposits are designed to encourage the return of containers, they may also (as is the case in Germany) be used as a sanction for the failure to achieve specific reuse/recycling targets.

Product/input charges These are charges levied on raw materials and/or intermediate inputs for products that are considered harmful to the environment, whether during the production process or during their consumption or disposal. In contrast to emission charges, which are associated with statutory sources, product charges are used for those cases in which consumption is not restricted to a particular area or number of users. The purpose of these charges is to reduce the use of these materials and to steer manufacturers to recycled or less harmful materials.

Disposal Taxes These are broadly applied, taxes related to waste management; their purpose is to make disposal more expensive, so as to make recovery or recycling more economically practicable. One should distinguish these taxes, which are applied directly to the cost of packaging, from taxes and fees for waste disposal (Demey et al., 1996). These taxes are generally supported by industry not only because they tend to be lower than other taxes, but also because they are broadly applied (no product/material discrimination). Moreover, in some countries (France), the taxes are used to fund environmental amelioration, while in others (the United Kingdom), they are used to reduce payroll taxes (Europen, 1997).

Issues raised by economic instruments

Economic instruments have come into favour in recent years because they permit companies to make the choice between paying for cost of the pollution or paying for the pollution control equipment; thus, if the cost of polluting becomes too high, the firm will cease its activity. The goal of the instruments is to internalise external costs that were hither to unpaid for, such as the cost of the pollution to the community or the loss of a resource. Unfortunately, the costing mechanisms are not yet well developed and as a result, some economic instruments have been quite arbitrary.

Some of the problems of economic instruments are caused by their variable nature, as there exist numerous factors which can affect the determination of which one is the best for that circumstance, i.e., available substitutes on the domestic market; substitutes which can be imported; available shelf-space; the consumer's reaction to a price increase to cover a deposit or tax; and of course, hygiene and product safety. Thus, it is nearly impossible to decide that one product, packaging or economic instrument is the best for all areas of a region or of a Member State. Moreover, economic instruments for packaging have traditionally focused on waste disposal instead of total environmental impact, i.e., the effect of disposal after use instead of the impact of the product on the environment throughout its life.

The nature of an eco-tax

The rationale of the eco-tax system is to affect the habits of consumers so as to reflect a greater respect for the environment. Thus, the government identifies those products that are harmful to the environment and then imposes an eco-tax sufficiently high that consumers begin to seek less expensive alternatives (which are, in theory, less harmful to the environment). The theory is that because of consumers' influence on the market, manufacturers would eventually begin to develop products that would be more environmentally friendly. The income from the taxes would eventually decrease as fewer of the more environmentally harmful products are produced (Price and Hauff, 1994). Yet, while it is true that the government usually collects the eco-tax without earmarking it, one should not assume that eco-taxes are a major source of revenue for the state; to the contrary, they often produce a negative result. For example, the maximum revenue obtainable from Belgium's eco-taxes was found to be 1.3 million ECU, whereas the cost of collecting the taxes was estimated to be at least 10.5 million ECU (Europen, 1997). Thus, eco-taxes should not always be considered as a guaranteed source of income.

The eco-tax system can also result in a form of 'double taxation' to the consumer (to whom additional taxes tend to be passed). For example, if batteries are packaged in plastic and cardboard wrap which has been made from bleached pulp, the product is then taxed twice under the eco-tax system (once for the content of the batteries and a second time for the content of the packaging), possibly rendering an essential product too expensive for many consumers. Batteries are also a questionable choice for an eco-tax because the theory of the eco-tax system is that it increases the cost of those products for which a more 'environmentally friendly alternative' is available. While batteries may not be very environmentally friendly, it is difficult to imagine what the ecological alternative to batteries might be. Moreover, a possible side effect of eco-taxes is that instead of products becoming more environmentally sensitive, only the more wealthy will be able to afford them.

Ironically, the cost of an eco-tax is often passed on to the consumer (as opposed to forcing the producer to pay for the cost of damage to the environment); this fact is not viewed as inconsistent on the grounds that the consumer may choose between taxed and non-taxed products (Demey et al., 1996). Yet, eco-taxes involve collection and administration charges which can make them quite expensive; the returns from them also diminish as firms find ways to qualify for the exemptions thereto. Moreover, there is no guarantee that the taxes paid will be used for the reparation of the environment. Lastly, recent economic studies have questioned the validity of the costs of eco-taxes imposed on packaging, as compared with other programs such as the Green Dot system (Europen, 1997).

The nature of a deposit system

Deposit systems are traditionally used to encourage the return of reusable materials such as wood packaging or refillable bottles. A 'voluntary deposit' is a system imposed by the manufacturer which reflects the cost of replacing the material if it is not returned. There are no set goals for rates of return. In contrast, a 'mandatory deposit' is one imposed by the state in order to achieve a policy goal; often the amount of the deposit has no relation to the actual cost of replacing the material. (It is important to note however, that deposits are not taxes because the unclaimed deposits are returned to the manufacturer, as opposed to the state.) Mandatory deposits are being considered in Greece, Austria, Germany and Ireland, for those instances in which the legislators believe that the recovery rates are insufficient. The goal of a mandatory system would be to push consumers toward refillable containers by raising the comparative price for non-refillables, and/or to foster the return of the container for refilling (Europen, 1997).

Deposit systems can be very effective in pushing consumers to return refillable containers, but they have several drawbacks: first, they may not be cost-efficient, often resulting in losses; and second, they create serious logistical and hygienic problems for retailers, who react by reducing the amount of shelf space for products with deposits, thus reducing the choices for consumers. More importantly, as the OECD has concluded, deposit systems can create barriers to trade where the amount of deposit is high compared to the cost of the product; when the logistics and expense of the system discourage the importation of goods; and when the deposit is only applied to certain materials or types of containers (OECD, 1993). However, one must note that for certain very harmful products such as waste oils, a mandatory deposit may be useful to assure proper disposal.

In the Community, one can also find a third, hybrid system, in which a 'quasi-voluntary' deposit is part of a range of recovery options permitted. Here, the operator still controls the deposit, but would not have instigated it without legislative intervention. Examples of this system would be the systems in France and Portugal, where the manufacturer can choose between setting up a deposit system or joining a recovery organisation, such as *Eco-Emballages*. In practice, manufacturers often make the latter choice. In Finland, in order to qualify for a lower packaging tax for obtaining a specified return rate, the manufacturer must also create a deposit system. In the Netherlands, Sweden and Denmark, deposit systems are required for PET bottles and certain types of metal cans.

Economic instruments and the Treaty

As with the reuse quotas discussed in Chapter 4, environmental instruments are of course, subject to the provisions of the Treaty. With eco-taxes as opposed to deposits, the arguments are a bit more interesting: while the ECJ made it clear in the *Danish Bottles* case that a deposit system is subject to the provisions of Article 30 et seq., it remains unclear as to what articles of the Treaty eco-taxes would be subject. Yet, in its 1997 Communication on 'Environmental Taxes and Charges in the Single Market',[2] the Commission itself noted that using eco-taxes to implement environmental policy naturally encroaches upon other Community legislation, such as that regarding the internal market, competition policy, taxation policy, and occasionally, agricultural policy.

Can Article 30 be applied to eco-taxes?

The provisions of Articles 30, 31 and 34 prohibit quantitative restrictions on imports or exports, or any measures having equivalent effect, and forbid Member States from introducing any new quantitative restrictions or measures which would have an effect equivalent thereto. The Commission has indicated that while Article 30 of the Treaty is not normally applicable to economic instruments, which are also subject to Articles 9, 12, 16 or 95 of the EC Treaty, it may be applied in certain circumstances. These would include: when the levy is imposed on an imported product where there is not a competitive or similar national product and the levy is so high that it interferes with Member State trade; where the implementation of the levy is separate from certain conditions thereof and these conditions (such as labelling requirements) would be subject to Article 30; and where the foregoing conditions result in a situation in which certain firms can affect the importation of the product in question. Thus, the Commission has concluded that the proportionality principle would only apply to the administrative portion and not the general principles of the eco-tax (Commission, 1997).

An interesting dispute has now arisen as to whether the actual levy itself (as opposed to its administrative provisions) could be subject to Article 30. As the Commission itself has noted, while there exists extensive case law regarding the application of Article 30 to non-fiscal legislation, there is no case law directly on point regarding the applicability of Article 30 to fiscal provisions. Nonetheless, certain case law of the ECJ has suggested that even fiscal measures could be subject to Article 30 and thus, is quite instructive. In Commission *v* Ireland,[3] commonly known as *Dundalk*, the Urban District Council of Dundalk had invited bids for a water supply contract, stating in the tender that the pipes must certify 'Irish standard 188'. The Council then refused to permit the bid of an undertaking using non-conforming pipes, despite the competitive price. The Commission alleged that the decision violated (among other legislation) Article 30 of the Treaty. The Irish government argued that Article 30 relates only to the sale of goods and that only Article 59 regarding services would be applicable, as the contract involved was for public works. The ECJ disagreed, holding that:

> Article 30 envisages the elimination of all measures which impede imports in intra-Community trade, whether the measures bear directly on the movement of imported goods or have the effect of indirectly impeding the marketing of goods from other Member States. The fact that some of those barriers must be considered in the light of specific provisions of the Treaty, *such as provisions of Article 95 relating to fiscal discrimination,* in no way detracts from the general character of the prohibitions laid down in Article 30 (para. 16, emphasis added).

Thus, the Court concluded that although other articles of the Treaty may address other matters, the application of Article 30 cannot be eliminated from the analysis. It is also interesting to note the opinion of the Advocate General, who declared that Article 30 should be applied in order to ensnare the maximum number of obstacles to the importation of goods, thus emphasising a broad approach to this article. Accordingly, although one could argue that Article 95 et seq. (discussed *infra*) is the only article that would apply to an eco-tax, the Court made it clear in *Dundalk* that if such a tax acts as a barrier to intra-Community trade, it must be analysed pursuant to Articles 30-36, even if it is a fiscal measure: 'The circumstance that certain impediments must be examined in the eyes of the specific provisions of the Treaty, such as those of Article 95 on tax discrimination, does not diminish in any way the general character of the prohibitions that make up Article 30'.

In any event, although in its Communication, the Commission cited case law contrary to this interpretation of *Dundalk,* all of this case law may be irrelevant in that none of these judgments were ever rendered following the enactment of directly applicable secondary legislation, such as the Directive. Thus, it is difficult to predict whether an economic instrument enacted pursuant to the Directive would be more strictly held to the rules of Article 30 et seq. Nonetheless, the fact that the ECJ has not yet addressed the issue of the applicability of Article 30 et seq. to fiscal measures does not mean that it is not ripe for discussion.

Moreover, it is worth noting that in the past, when Member States used only command-and-control instruments, they were always bound by the provisions of Article 30. However, as some Member States have begun implementing eco-taxes, they have argued that Article 30 no longer applies and that Articles 90-95 give them *carte blanche* to do as they please. Yet, the change from more traditional instruments to eco-taxes should not permit Member States to thwart the Treaty's fundamental premise of the free movement of goods. Furthermore, it is nonsensical that Article 30 et seq. would not apply to eco-taxes when it does apply to other economic instruments, such as deposits.

Application of Article 30 et seq.

On the other hand, even if one assumes that Article 30 et seq. does apply to an eco-tax, the tax (or other form of economic instrument) may nonetheless be valid if it complies with these articles. One must first ask whether the economic instrument constitutes a quantitative restriction on imports or an equivalent effect thereto. Although an economic instrument does not openly restrict imports, its effect (or potential effect) could be to reduce the number of imports to the Member State imposing it, due to the financial and logistical costs associated therewith. Thus, a strict reading of Article 30 would prohibit an economic instrument if such an effect or potential effect on imports could be demonstrated.

Exemption pursuant to Article 36

Assuming then that the economic instrument would constitute a quantitative restriction, one must next consider whether it would be exempted pursuant to Article 36 because its objective would be to protect the environment. This objective may include a preventative aspect or concepts such as the polluter-pays principle or the precautionary principle (Commission, 1997). However, as discussed more fully in Chapter 4, in order to qualify for such an exemption, it is not sufficient that the economic instrument have as its objective the protection of the environment: the economic instrument must additionally not serve as a method of arbitrary discrimination or a disguised restriction on trade between the Member States.

Method of arbitrary discrimination or a disguised restriction on trade?

With respect to an eco-tax, it has been argued that such a tax would not raise any trade barriers because products manufactured in the taxing Member State are not given more favourable treatment by being taxes at a lesser rate than those imported from other Member States. However, in certain Member

States such as Belgium, some provisions of the eco-tax were determined in consultation with the Belgian paper industry regarding its production methods: as a result, most paper manufacturers in Belgium qualified for a reduction, if not an exemption from the eco-tax, while the non-Belgium paper manufacturers had to modify their production procedures in order to avoid taxation. Similar arguments can be made regarding beverage containers made from PET versus PVC (Price and Hauff, 1994). It is clear then that the legislator must be careful that the drafting of such economic instruments remains transparent. Otherwise, a presumption of arbitrary discrimination will arise.

Similarly, a deposit system may be a method of arbitrary discrimination where the finances and logistics required for such a system places a heavier burden on foreign packaging. As the ECJ noted in *Danish Bottles,* although such a system may be permissible in order to protect the environment, it must be applied indiscriminately to domestic and imported goods and must be proportionate to the goal sought, in that the deposit required constitutes the minimum restriction possible to the free movement of goods.

Articles 9-12

Pursuant to Articles 9-12, a Member State may not introduce any new customs duties on imports or exports, or any charges having equivalent effect. While in the strict sense of the word, it is true that these 'customs duties' no longer exist, ECJ case law has made it clear that any charges imposed on products only because they cross a border would have an effect equivalent to a customs duty (Demey et al., 1996). Thus, Articles 9-12 would forbid such a charge where it is imposed on foreign goods. However, if the levy is part of an overall system of taxation, it would have to be examined pursuant to Article 95 et seq. (Commission, 1997).

Another example of a levy having an effect equivalent to a customs duty would be where a large part of the domestic manufacturers are exempted from a general system of taxation. In this regard, the ECJ has suggested that it may become necessary to consider how the revenue from the tax is utilised. For example, even if the charge is applied equally to domestic and foreign products, if its revenue is used to compensate domestic producers, it constitutes a charge having equivalent effect. Moreover, if the revenue is given to the *consumers* of the products, the charge has to be judged under the state aid rules[4] as well. With regard to state aid, Article 92 provides that any aid granted by a Member State which either distorts or threatens to distort, competition in the Community, shall be declared invalid. According to the general rule, even an aid which comprises part of a levy will be invalid if it is incompatible with other provisions of the Treaty, notably, Articles 9-12, 30-36 and 95 (Commission, 1997).

Article 95

Article 95 forbids a Member State from imposing a tax (whether directly or indirectly) on the product of another Member State, where said tax exceeds that imposed on a domestic product, or where the tax gives indirect protection to other products. The goal of Article 95 is to assure that national tax regulations remain neutral; it thus forbids both the use of the regulations to discriminate against products from other Member States and the use of national taxation schemes to protect domestic products. The jurisprudence of the ECJ in this area is based on three canons.

The first canon assumes that as a general principle, Member States are permitted complete liberty to choose the taxation system they wish. According, the ECJ has held that Member States *can* have differing levies between national and foreign products, even if the products are 'similar'. However, such differentiation must: 1) be founded on an objective standard, such as the type of raw material or the process; 2) have as its purpose the attainment of an economic policy which is compatible with all Community legislation; and 3) the guidelines of the differentiation must not discriminate against imports from other Member States or protect national products (Commission, 1997). One should note that while the first two criteria can be objectively identified, the third criterion is quite subjective and would have to be examined on a case by case basis. Here, the burden of proof would be carried by the Member State imposing the tax (Demey et al., 1996). Interestingly, the jurisprudence of the ECJ has indicated that an eco-tax is not necessarily discriminatory simply because it falls heaviest on a product from another Member State; however, the Member State imposing the tax must demonstrate that the foregoing is the result of objective, non-discriminatory criteria (Commission, 1997).

The second canon is that, when determining whether products from outside a Member State are taxed more heavily, the point of reference will be the taxation system of the Member State in question. However, this system must be sufficiently transparent that one can determine if the criteria used are truly objective. In analysing the system, one must consider: '1) the rate of the levy; 2) the terms regarding the taxable base; 3) the control systems for charging the levy; and 4) the rules for the collection of the levy'.[5] Moreover, the fact that a tax is imposed chiefly on goods from other Member States is not sufficient proof that it is discriminatory. The determining factor will be the true cost imposed on the domestic product as compared with the import. Yet, as the Commission has noted, 'Article 95 does not give the Community a right to judge whether a levy in a Member State is excessively high in relation to its environmental objective'.[6]

The third canon is that Article 95 would be infringed if a product from outside a Member State is charged at a higher rate than a domestic product. The determinant of this canon is the 'product' itself, i.e., the similarity among both the domestic and foreign products taxed. The ECJ has ruled that similar products are those that 'have similar characteristics and meet the same needs from the point of view of consumers'.[7] The Commission has indicated that one must also consider whether products that have the same function, but have different environmental characteristics because of their differing content or processes can be considered as different products (Commission, 1997).

Overall, the ECJ's varied jurisprudence suggest that the identification of a discriminatory eco-tax often has to be made on a case-by-case basis. For example, an eco-tax that works in favour of a product, of which the majority is manufactured nationally, would be discriminatory. Another discriminatory eco-tax would be where most domestic goods are able to avoid the tax whereas it is successfully imposed on imports. Lastly, an interesting situation can occur where an eco-tax is only imposed on imported goods because no similar products are manufactured nationally as the direct result of the eco-tax; in this instance, the tax would be discriminatory (Demey et al., 1996).

It is also important to note that when the imported products are different, but continue to compete with the national products, the tax imposed must not protect the latter. One must consider both the actual and potential market that would exist for imported products in the absence of a protectionist tax. Moreover, *how* the revenue from the eco-tax is used can be an additional important factor. For example, if the revenue is used to offset the tax imposed on domestic products, this would be discriminatory. Thus, an eco-tax could be considered as both discriminatory and inappropriate state aid, because it is subject to both Article 95 and Article 92.

Economic instruments and the Directive

In addition to being subject to the rules and case law of the Treaty, economic instruments are of course, subject to all other provisions of the Directive. In particular, Articles 2, 18 and 23 contain possibilities for conflict.

Articles 2 and 23

As may be recalled from Chapter 4, Articles 2 and 23 address the types of packaging to which the Directive applies. Article 23 specifically repeals

Directive 85/339/EEC regarding containers of liquids for human consumption, while Article 2 indicates that the scope of the Directive encompasses all packaging that is placed on the market. Accordingly, it may be argued that economic instruments which target a particular type of packaging (such as beverage containers) or a particular type of material (such as plastic) violate these articles. It has been suggested that the Directive does not outright prohibit specific measures being taken for a specific sector and that because the Directive sets overall objectives and provides Member States latitude to achieve them, a Member State may implement different measures 'in parallel, which do not necessarily have to be similar'. However, such distinctions should be supported by objective, scientific reasoning; it is not clear that the economic instruments imposed on packaging to date have done so.

Article 18

Article 18 provides that as long as a package complies with the provisions of the Directive, a Member State shall not create an obstacle to such a package being placed in its territory. Thus for example, while an eco-tax on packaging which exceeds the concentrations of heavy metals permitted under the Directive would be acceptable, whereas an eco-tax on a type of packaging which is permitted under the Directive should not. It has been argued however, that an eco-tax or deposit system could be appropriately imposed on a material that is acceptable pursuant to the Directive, but which is so difficlut to recycle in a particular Member State that the Member State will not reach its target recycling rate of 15% for that material (Demy et al., 1996). Yet in light of Article 18, it is clear that Member State economic instruments imposed on specific types of packaging or on certain packaging materials that are acceptable pursuant to the Directive and/or that are circulating in other Member States can only be justified on the basis of objective, scientific evidence. Thus, the goal of some Member States to use economic instruments without such support to eliminate or to sharply reduce the market share of certain types of packaging would be in clear violation of these articles.

Economic instruments and distortions of competition

It is clear that when certain products and industries are subject to eco-taxes, distortion of competition may occur. For example, the Belgian newspaper publishers have strongly protested the eco-tax on paper because it applies to paper used for newspapers, but not paper utilised for magazines; thus, news magazines will have a competitive advantage. Claims have also been made that the eco-tax system and its administrative body constitute a public undertaking for purposes of Article 90 of the Treaty and are infringing the Treaty because the eco-taxes distort competition among entities in different Member States (Price and Hauff, 1994).

Although the ECJ has not rendered any judgments regarding economic instruments and packaging waste *per se,* the recent case of BIC Benelux *v* Belgium[8] is instructive as to the Court's concern regarding the effect of economic instruments on the internal market. Here, BIC Benelux, which sells disposable razors in Belgium, alleged that a Belgian eco-tax on its razors was invalid because Belgium had failed to notify the economic instrument to the Commission, in violation of Directive 83/189/EC regarding technical specifications. Belgium asserted that notification was unnecessary, as the tax fell within the exemption for fiscal measures in the directive. However, the Court ruled that while the fiscal aspects of the measure would be covered by exemption, the requirement to affix special markings to the products constituted a technical measure and therefore must be notified to the Commission. Although at first blush, this case appears to be purely an issue of technical standards, it also reflects the Court's opinion that where technical specifications are part of a measure, they must always be examined for their effect on the free movement of goods, whether they are a part of a technical standard itself or a part of some other type of measure that would normally not be notified to the Community. Thus, even fiscal measures which contain a partial technical specification cannot be exempted from review, as they could still impact upon the free movement of goods.

In contrast, with regard to 'pure' economic instruments, Member States appear to be given free rein to do as they please, even if they create obstacles to trade! It has thus been remarked that 'the freedom allowed the Member States in direct and indirect taxation matters is obviously out of synch with the market realities'.[9] Given the increasing use of economic instruments by the Member States, it is likely that more guidelines from the Commission and/or jurisprudence from the ECJ will be forthcoming on this issue in the near future.

Types of economic instruments implemented in certain Member States

Belgium

The Belgian government decided not to wait for the Directive to be implemented and instead, created its own eco-tax program to reduce production (and therefore consumption) of certain packaging which allegedly posed a threat to the environment. The Belgian eco-tax system is designed to impose taxes on (and thus increase the costs of) those products for which a more environmentally friendly alternative is available. The entity responsible for paying the eco-tax is the person or company that introduced the product onto the Belgian market. Interestingly, the original concept was to have been royalties system, similar to that of Germany, and in fact, the Fost Plus system was created by Belgian industry. However, in 1993, the government decided that an eco-tax system would be implemented instead. Although the original plan was for the eco-tax system to begin in July of 1993, the start date was postponed to 1 January of 1994, due to various political and economic concerns. The first items to be affected were beverage containers for sparkling water, beer, colas or lemonade; containers made of PVC were added on 1 July 1994. All other beverage containers became subject to the tax at 31 December 1997. (Dallemagne et al., 1993; Price and Hauff, 1994).

With regard to beverage containers, the Belgian tax is imposed regardless of the container's size or composition. A container will be exempt if it can be reused at least seven times. Beverage containers which are non-refillable may only be exempted from the tax where certain annual recycling rates are met. While recycling may be achieved by the packaging producer itself, in reality the Fost Plus programme is responsible for the recycling needs of all the major beverage manufacturers (ERRA, 1998a).

The entity placing the packaging/product on the market must pay the eco-tax; a registration system has been established for this purpose. In addition, special packaging markings are required, which indicate: 1) if an eco-tax must be paid and if so, how much; 2) if the packaging is exempt from the eco-tax for the aforementioned reasons; or 3) if the packaging is exempt because it fits the reusability requirements (ERRA, 1998a).

Moreover, it should be noted that the eco-tax system in Belgium does not merely apply to packaging materials. Products themselves, such as batteries, pesticides, phytopharmaceutical products and disposable products for non-medical uses (for example, razors and disposable cameras) are also subject to the eco-taxes. The Belgian eco-tax system is regulated by a commission that studies not only the effect of the system on the environment, but the effect of the system on unemployment as well. The commission's power includes the ability to create new levels of recycling, or even eliminate eco-taxes where necessary to protect employment levels. In fact, if employment levels come under threat, eco-taxes may be reduced or eliminated altogether for certain categories of materials (Price and Hauff, 1994; ERRA, 1998a).

In December 1997, the Commission issued an infringement letter which stated that Belgium had not properly implemented the Directive. Belgium responded by notifying the Commission in February of 1998 of its legislation regarding the eco-taxes, as required by the Directive. There have been several discussions among trade associations regarding a challenge to some of the eco-tax provisions (ERRA, 1998a). Such a challenge could be successful because as discussed *supra,* should be Belgian eco-taxes be subject to Article 30 et seq., they would likely be held discriminatory.

Germany

In Germany, several municipalities have tried to impose an eco-tax at the local level. In July of 1992, the city of Kassel imposed an eco-tax on 'catering packaging'. As a result, taxes were imposed on plastic and paper crockery, breakfast packs for condiments and on plastic cutlery. Several restaurants and associations immediately filed suit, arguing that the tax was a *de facto* ban on catering packaging and also resulted in a dual taxation under the German packaging legislation which implemented the Directive. In May of 1998, the *Bundesverfassungsgericht* (Federal Constitutional Court) ruled that municipal taxes on such packaging are unconstitutional and counter to the objectives of the German Packaging Ordinance, as such taxes only affect part of the packaging chain (ERRA, 1998a). This ruling is interesting when one considers that, like the catering tax, the German quota on refillables only affects one part of the packaging chain. One could however, distinguish the catering tax on the ground that because there was no exemption clause, the tax effectively resulted in a ban on the goods taxed, whereas if the quota on refillables is not met, the deposit system then applies. However, one can then counter-argue that the deposit system is simply another form of a trade barrier, as it would have the same effect on imported goods.

One should also not that in any event, the concept of the catering tax was questionable, as one intensive study has suggested that reusables in a fast-food restaurant can only result in lower emissions and energy consumption (as compared to disposables) if they are used several hundred times.[10] Moreover, other studies have indicated that reusable cutlery and crockery have a lower hygiene rate than disposables, possibly because space considerations in smaller facilities result in less hygienic methods being used (Europen, 1997).

Denmark

In early 1998, Denmark imposed a new system of increased taxes on all beverage containers, including glass and plastic bottles; cardboard and laminates; and certain metal cans (ERRA, 1998a). The only beverages exempt from these taxes are milk products, which raises questions of discrimination against foreign packaging, as the quality issue of milk products tends to result in their being of a domestic nature.

Sweden

While Sweden has not imposed eco-taxes *per se* on packaging, it does have a system of reimbursable deposits on beverage containers; this system was first implemented in 1973. As with Denmark, although deposits are not imposed on containers for milk products, this has not been challenged by industry. In addition to its deposit system, Sweden has established a permit system for all entities wishing to place aluminium or PET containers on the market. These permits can only be obtained after the applicant proves that certain conditions can be met, such as the imposition of a deposit; specific labelling requirements; and a recovery rate of 90% (ERRA, 1998a).

The Netherlands

Due to the success of its packaging covenants, the Netherlands did not implement any taxes on packaging until 1994. The tax implemented was supposed to end in 1996, but it continued for beverage cartons in order to encourage recycling. The Netherlands also has a deposit system for refillable and non-refillable PET bottles (ERRA, 1998a).

Finland

In 1994, Finland enacted legislation which implemented two economic instruments. The first instrument imposed an eco-tax on all containers for beer and soft drinks, depending upon whether the container was recoverable. However, following pressure from industry, the eco-tax was reduced, on condition of creation of the second instrument, a deposit scheme for all beverage containers. There do not appear to be any exemptions to this tax and industry has strongly opposed it, but without success (EIS, 1996b; ERRA, 1998a). Moreover, the deposit system, which varies the deposit according to the type of packaging material, is certainly an infringement of Articles 2, 18 and 23 of the Directive, as discussed *supra*.

Luxembourg

In 1995, when Luxembourg notified a draft law implementing the Directive, the Commission responded by issuing a Detailed Opinion. The law would have required an eco-tax to be paid if recycling and refilling targets for beverage containers could not be met. The Commission stated that the law would discriminate against imported packaging and was disproportionate, yet Luxembourg had failed to show scientific evidence that refilling was better than recycling. Luxembourg had relied on a German study, but the Commission rejected this argument for two reasons: first, the situation in

Germany is quite different from that of Luxembourg and the study is not transferable. Second, the German study indicated that refillables were only better under certain conditions, i.e., for short distances. The Commission concluded that the issue had to be decided on a case by case basis (Europen, 1997). One must note however, that this situation does not seem much different from that of the German refill quota and indeed, it appears that Luxembourg was hoping to copy the successful German reuse quota.

Environmental agreements

References to environmental agreements in the Directive

References in the Directive to 'voluntary agreements' are limited to one paragraph in the definitions section, which was added at the request of several Member State:

> 'voluntary agreement' shall mean the formal agreement concluded between the competent public authorities of the Member State and the economic sectors concerned, which has to be open to all partners who wish to meet the conditions of the agreement with a view to working towards the objectives of this Directive.

Although the definition was added primarily in recognition of packaging agreements in the Netherlands, Belgium and Denmark, since the enactment of the Directive, other Member States have also begun to use this instrument in partial implementation of the Directive.

Environmental agreements in general

Environmental agreements (or 'covenants', as they are known in the Netherlands) may be described as instruments, typically contracts which are created via consensus between government and industry (and which, more recently, involve the participation of NGOs and other third parties). While environmental agreements have most recently been used to achieve environmental objectives, they may be concluded for other purposes as well (Koeman, 1993). Their common theme however is that they are all voluntarily concluded and that participation in them is voluntary as well.

There are two types of environmental agreements one sees with respect to packaging. The first type are 'preparatory' environmental agreements, which are designed to create rules before legislation has been enacted, or sometimes in anticipation of legislation which will be quite strict: an example of this type would be the Dutch packaging covenant, which was enacted in 1991, following the implementation of strict legislation in Germany. The second type of environmental agreement is an implementation contract designed to achieve the goals of the Directive; an example of this type would be the agreement in Finland regarding increased recycling.

Environmental agreements and the Treaty

The Treaty does not contain any specific provisions regarding environmental agreements, as at the time it was enacted, this type of contract had not been created. Environmental agreements are however, considered as valid law in the Community, as evidenced by a recent communication from the Commission on this subject.[11] Yet, one area with which environmental agreements may come into conflict is that of competition policy.

Article 85 of the EC Treaty Pursuant to Article 85, it is prohibited for any environmental agreement to which a firm or trade association is a party, to have as its 'object or effect, the prevention, restriction or distortion of competition within the common market'. However, Article 85(3) of the Treaty provides exemptions where the restriction either contributes to the improvement of the production or distribution of goods, or promotes technical or economic progress. Nonetheless, these restrictions must result in consumers being given a fair share of the benefit that results. Furthermore, the restrictions: 1) must be indispensable to the attainment of the objectives of the agreement; and 2) must not allow the firms to eliminate competition with regard to a substantial part of the products affected by the environmental agreement (Commission, 1996).

The question is thus presented as to whether the protection of the environment would qualify as one of the exemptions under Article 85(3). The Commission suggests that it would when it states that, 'the protection of the environment might be considered as an element which contributes to improving the production or distribution of goods and to promoting technical and economic progress' (Commission, 1996). According to the Commission, it would apply the proportionality principle and weigh the restrictions of competition which would ensue from the environmental agreement, against the value of its environmental goals. Put another way, the Commission examines the restrictions of the environmental agreement to determine if they are truly necessary and compares this with whether consumers would in point of fact receive their fair share of the benefits arising from the achievement of its goals.

The very nature of environmental agreements requires that competing firms participate in discourse whereby they exchange information. Indeed, a fundamental aspect of all voluntary approaches is the co-operation among the affected firms to commit to the approach and to exchange information regarding technology. Unfortunately, this same co-operation raises questions of the reduction of free-market competition and the possible introduction of non-tariff barriers. But while environmental agreements often require businesses to reach a consensus amongst themselves, it remains difficult to motivate firms, as they are naturally reluctant to work with their competitors unless there is a greater threat or external benefit to be gained. In the case of packaging, this threat is the underlying legislation, which often imposes traditional command-and-control legislation, including penalties, in the event of non-compliance.

Notwithstanding the fact that an environmental agreement could arguably fall within an exception of Article 85(3), it must be notified to the Commission for its review and comment. Procedurally, the environmental agreement must be notified to DG IV of the Commission for its review before the agreement is implemented, as the Commission has the exclusive power to grant an exemption under Article 85(3). As environmental agreements cannot create barriers to the internal market, agreements for products which contain any restrictions on the free movement of the goods must base these restrictions on health or environmental grounds. Naturally, these restrictions must be proportional and not arbitrary and must not be a disguised restriction on trade within the Community. When environmental agreements contain technical specifications for products, said specifications must be first screened by the Commission and then communicated by the Commission to the other Member States. (This requirement also applies to all environmental agreements to which a public body is a party). Both the Commission and the Member States are given a time period in which to reply and comment on the proposed specifications (Commission, 1996).

Article 86 of the EC Treaty An environmental agreement is also bound by the provisions of Article 86 that it may not lead to an abuse of a dominant position within the internal market to the extent that it affects trade between the Member States. Within a Member State, firms in the packaging industry may work together to implement the Directive. It is here that the possibilities for non-tariff barriers could arise because environmental agreements at the Member State level are often promoted by and receive support from, national departments of industry (Lévêque, 1997a). If one considers some of the national waste recycling programmes (which are in essence, environmental agreements) such as *Eco-Emballages* in France or *Duales System Deutschland* in Germany, one can see that the collective action aims at or can result in, the freezing of competition between firms with respect to environmental performance. For example, there are certain materials such as aluminium or paper that have a better environmental performance than other materials. But when the discussions regarding the Directive began, it was decided not to differentiate among the materials used. Thus, the competition and innovation among firms regarding the identification and use of the most environmentally-friendly materials has been frozen. The result is that market shares of the material have not changed significantly, inasmuch as the use of the materials did not change because there was no incentive to develop more environmentally-friendly materials (Lévêque, 1997b).

If the environmental agreement provides that a private party is to perform some type of public activity which could restrict competition, then Article 86 would apply. One should note that while Article 86 does not contain any exemptions, it would simply not be applied in the event the Commission decides not to pursue a matter. For example, to date, the Commission has chosen not to pursue allegations of an abuse of dominant position regarding the *Eco-Emballages* or *Duales System Deutschland* packaging waste programmes. On a different level, in an oligopolistic market, an environmental agreement among firms (regarding for instance, the use of a particular technology) could arguably constitute an abuse of dominant position, as prohibited by Article 86.

The types of environmental agreements employed by Member States

As noted in Chapter 3, several Member States utilised environmental agreements to implement the 1985 Liquid Foodstuffs Directive; the majority of these were however, highly criticised. The new environmental agreements that have been executed to implement the Directive have taken some of the problems of the older agreements into account.

The Netherlands The most well-known of environmental agreements regarding packaging is the 1991 Packaging Covenant in the Netherlands, which was presented as a successful compromise between industry and public authorities. Although there was not officially any legislation threatened, the covenant provided a popular alternative to the packaging waste ordinance which had recently been enacted in Germany and which was the subject of negative publicity. However, since that time, serious problems have arose with free riders and it became necessary to enact underlying legislation (Koeman, 1997). In December of 1997, a new Packaging Covenant II (for the years 1998-2001) was executed; it consists of an umbrella covenant and six sub-covenants for producers and importers; glass; paper fibre; metal; plastics; and wood. This Covenant helps implement the Waste Law of 1993, which contains a strict hierarchy of: prevention; product reuse; material recycling; incineration (first with, and then without, energy recovery; and as a last resort, controlled landfilling. While the Packaging

Waste Regulation of August 1997 requires firms to meet the maximum targets of the Directive more than two years earlier than required by thereby, firms are exempt from these requirements if they join Covenant II, which gives them until 2001 to meet the overall target of 65% (indirect recycling targets vary according to the material) (ERRA, 1998a).

Germany An example of an 'implementing' environmental agreement would be the contract to create the system *Duales System Deutschland,* which was established to implement the 1991 Packaging Waste Ordinance. Until recently, this programme had severe financial problems as a result of the excessive plastic waste collected. Further problems arose arise from the classic free rider problem. As noted in Chapter 4 however, the system has ultimately been vindicated as an cost-efficient and effective method for the handling of packaging waste.

Denmark In contrast to other Member States, Denmark has had numerous environmental agreements regarding packaging, such as its 1992 agreement for the recycling of 80% of non-refillable glass. Other agreements include a 1991 accord regarding the use of PET containers and an agreement in 1994 for the recovery of transport packaging (Europen, 1998b).

Finland This Member State has one environmental agreement implementing the Directive, a 1995 agreement to achieve increased recycling and reuse of packaging and packaging materials. This agreement addresses Finland's unusual demographic structure (in which the majority of the population is concentrated in southern Finland) by empowering municipal authorities to conduct the tasks of collection, transportation and implementation of appropriate levies where it would not be either economically or environmentally beneficial to conduct these tasks separately (ERRA, 1998a).

Belgium Lastly, several environmental agreements were enacted in Belgium prior to the enactment of the Directive. An initial agreement in Flanders in 1991 was followed by other agreements for both the Brussels region and Wallonia in 1992. Moreover, in 1997, an Interregional Agreement was executed, which provided for a decree for each region in order to implement the Directive. This Interregional Agreement has served as the framework for the implementation of the Directive by specifying requirements for the prevention of packaging waste; the selection of targets for recycling and recovery (which exceed those of the Directive and thus, had to be notified to the Commission); the imposition of a take-back obligation for those placing packaging on the Belgian market; as well as the provision of information to the Belgian Packaging Commission.

Conclusions

In recent years, the frustrations of Member States regarding the enforcement of command-and-control legislation has led them to seek alternative methods of implementing all types of environmental legislation. With regard to packaging waste, Member States have been particularly fond of using economic instruments and environmental agreements, both to implement the Directive and to achieve their policy goals. However, as these instruments were not envisaged at the time the Treaty was drafted, they must be closely scrutinised to assure that they do not violate its fundamental principles. Although the use of economic instruments and environmental agreements has not led to the same degree of division within the Commission as has been caused by the implementation of reuse quotas, the final word regarding infringements of the Treaty and/or the Directive will probably again be left to the Court of Justice.

Notes

1 Amended proposal for a Council Directive on packaging and packaging waste, Official Journal No. C 285, 21 October 1993.
2 Communication from the Commission, 'Environmental Taxes and Charges in the Single Market', COM (97) 9 Final, 26 March 1997.
3 Commission of the European Communities *v* Ireland, Case 45/87 [1988] ECR 492.
4 See Chapter 6 for a further discussion of state aids.
5 Commission, 1997, p. 11.
6 Commission, 1997, p. 8, citing Commission *v* Greece [1990] ECR 1567.
7 Commission, 1997, p. 9, quoting Commission *v* Italy [1987] ECR 2013.
8 BIC Benelux *v* Belgium, Case 13/96 of 20 March 1997.
9 Jean-Pierre Hannequart, quoted in EIS, 1995.
10 Porter & Perchard, 'Environmental and public health aspects of reusable and disposable food service packaging' (1996).
11 'The Commission's Communication on Environmental Agreements', 27 November 1996, COM (96) 561 final.

6 Packaging Legislation and Competition Policy

Introduction

As we have seen in Chapters 4 and 5, varying interpretations of how to implement the Directive can lead to conflicts among Member States and among participants in the packaging chain. While the foregoing chapters addressed issues which were particular to one or two Member States, there are other packaging waste issues which are Community-wide in scope. Chapter 6 considers competition issues raised by the Treaty and the Directive, as well as the use of commercial markings at the European level and the effects of state aids; it concludes with a look at general competition policy issues.

Competition issues raised by the Treaty and the Directive

Articles 85 and 86 - the competition rules

Articles 85 and 86 of the Treaty set forth the competition rules for the Community. Article 85(1) prohibits 'all agreements between undertakings, decisions by associations of undertakings and concerted practices which may affect trade between Member States and which have as their object or effect the prevention, restriction or distortion of competition within the common market...', and then provides a list of examples of restricted behaviour. Article 85(2) states that such agreements shall be void, while Article 85(3) identifies those circumstances under which the Commission may grant an exemption. In general, Article 86 prohibits 'any abuse by one or more undertakings of a dominant position within the common market or in a substantial part of it... in so far as it may affect trade between Member States', and then sets out a non-inclusive list of specific examples.

Application of Article 90

Although it is clear that the competition rules would apply to the environmental agreements for waste management discussed in Chapter 5, the question is presented as to whether the national waste treatment associations (such as *Duales System Deutschland* or *Eco-Emballages*) that are mandated by Member States to collect and sort waste in order to comply with the Directive, would also be subject to these rules. The answer lies within Article 90(1), which provides that:

> In the case of public undertakings and undertakings to which Member States grant *special or exclusive rights,* Member States shall neither enact nor maintain in force any measure contrary to the rules contained in this Treaty, in particular to those rules provided for in Article 6 [nationality] and Articles 85 to 94 [competition and state aids] (emphasis added).

It would thus appear that Article 90 has the effect of applying the provisions of Articles 85 and 86 to private enterprises which are given quasi-public powers. However, Article 90(2) contains an exception:

> Undertakings entrusted with the operation of services of general economic interest or having the character of a revenue-producing monopoly shall be subject to the rules contained in the Treaty, in particular to the rules on competition, *in so far as the application of such rules does not obstruct the performance, in law or in fact, of the particular tasks assigned to them.* The development of trade must not be affected to such an extent as would be contrary to the interests of the Community (emphasis added).

It has been suggested that this article would not apply to monopolies granted by Member States for the collection of wastes because such an application would prevent the undertaking from completing its assigned task (de Sadeleer, 1995a). However, it is hard to visualise how applying the competition rules so that other waste treatment operators can enter the market would prevent an organisation such as *DSD* or *Eco-Emballages* from continuing its mandate. In fact, given the recent move toward globalisation of waste treatment (Gotschall, 1996), it would appear that greater efficiencies and lower costs could be achieved by developing competitive bidding for packaging waste treatment.

It is more likely that Article 90(2) was meant to apply to a service that cannot be properly provided by other undertakings in the market (such as, until recently, the telecommunications sector). Indeed, Article 90(2) has been narrowly interpreted, with the ECJ even holding that the classification of the undertaking and/or the legal framework within the agreement or decision was made, as either public or private made no difference to the application of competition rules (Steiner, 1996). Thus, it does appear that waste operators mandated within the framework of the Directive would be subject to the provisions of Articles 85 and 86.

Moreover, although the Commission has thus far not brought any proceedings against the national waste treatment organisations, it did render an instructive opinion in the context of a complaint by Spa, the Belgian mineral water producer. Spa had complained that the German market for mineral water was inaccessible because of: 1) the German reuse quotas; and 2) the fact that it was impossible to affiliate with *Genossenschaft Deutscher Brunnen* ('*GDB*'), a company which treated reusable mineral water bottles in Germany. The Commission opened an investigation, in the course of which it made several interesting observations. Although the Commission did not comment on the German Ordinance, it did conclude that *GDB* had created a barrier to trade against foreign producers wishing to enter the German market, within the context of Articles 85 and 86. The Commission further concluded that, given *GDB*'s dominant position in the market for mineral water packaging, its refusal to grant access to foreign producers was an abuse of dominant position, pursuant to Article 86. Following these comments and further negotiations, the matter was closed after *GDB* agreed to open up its pooling agreement to non-German firms (de Sadeleer, 1995a). It is thus clear that Articles 85 and 86 can be applied to associations mandated to collect, sort and treat packaging waste.

Application of Articles 85 and 86

With respect to Article 85, agreements regarding waste treatment can be both horizontal and vertical. Horizontal agreements would include an agreement between all entities that produce a specific type of waste, while vertical agreements would be between the packaging manufacturers; the packaging distributors; the collection and sorting firms; and the entities which treat packaging waste (de Sadeleer, 1995a). A hypothetical example of such an agreement would be where legislation that is enacted by a Member State delegates powers to a private association, which later results in an unfair restriction of competition. In this case, there would be two separate infringements: first, the Member State would be in violation of Articles 3(g) and 5 of the Treaty and then the association would be in violation of Article 85 and/or Article 86 of the Treaty (Commission, 1996).

In order to obtain an exemption under Article 85(3), the agreement or concerted action must contribute 'to improving the production or distribution of goods or to promoting technical or economic progress', although it is not necessary that the agreement meet all four of these criteria. The issue then arises as to whether the protection of the environment would constitute one of these criteria. It is true that the Commission has granted some exemptions for environmental purposes, but these matters are usually determined on a case-by-case basis. Moreover, it is important to note that in order to obtain an exemption, the agreement must be notified to the Commission *in advance* and not be the subject of an investigation (Steiner, 1990). Accordingly, it is not completely clear that an environmental purpose will be sufficient to obtain an exemption under Article 85(3).

If we turn to Article 86, it is clear that a mandated association such as *Duales System Deutschland (DSD)* or *Eco-Emballages* is an undertaking in a dominant position; the question is whether its conduct constitutes an abuse. Arguably, the requirement that manufacturers who wish to sell their products in the Member State become members of the national waste treatment system would be an imposition of an unfair trading condition, as forbidden by Article 86. One could also argue that Article 86 would be infringed because the restriction of the Green Dot limit the national market to those producers and importers who comply with its regulations; as a result, consumers are prejudiced because their choices of products are restricted. Of course, the defence of a national association would be that manufacturers do not have to join its programme, as they can assume individual responsibility for taking back their packaging. But the realities of the marketplace and the near impossibility to stock a product at major retailers has effectively resulted in the Green Dot being required for all manufacturers in those Member States where it is used. With regard to possible defences to an Article 86 action, it is important to also note that unlike Article 85, Article 86 contains no exemptions, whether for an environmental purpose or otherwise. Thus, should the Commission decide that a quasi-public association or an environmental agreement is infringing Article 86, the fact that the activity was conducted for an environmental purpose would not be a valid defence.

Article 7.1 of the Directive

One of the issues raised by the Directive is the fact that in each Member State, a *de facto* monopoly has been granted to a private undertaking in the area of waste management. For example, while there are several organisations in France responsible for collection, sorting and recovery (*Adelphe* for liquor and wine bottles and *Cyclamed* for packaging materials from the pharmaceuticals industry), the main organisation, *Eco-Emballages,* is by far the largest. The assumption has always been that it would be impossible to have more than one firm responsible for the majority of packaging materials; this has had the effect of stifling competition.

Article 7.1 of the Directive, in providing for the creation of return, collection and recovery systems, states that 'these systems shall be open to the participation of economic operators of the sectors concerned and to the participation of competent public authorities'. Article 3.11 of the Directive defines 'economic operators' to include *inter alia,* 'authorities and statutory organisations'. It is thus clear that a Member State can license more than one packaging waste association. In an ironic twist, given that it was the first major packaging waste association in Europe, Germany's *DSD* has recently begun to face competition from other German waste management companies operating at regional level, while in the United Kingdom, several different waste treatment programmes now compete for business (ERRA, 1998).

What is also evident is that in the past few years, waste treatment (from collection and sorting to recuperation of energy and/or landfilling) has gone global. Inter-country alliances among waste operators have occurred, but other than the co-ordination of the Green Dot, none have yet taken place in the domain of packaging waste treatment (Gotschall, 1996). Accordingly, the mandates granted by Member States for the collection, sorting and treatment of packaging waste should be opened up to competitive bids. Indeed, one can envisage that economies of scale may be more easily achieved by a firm operating in several Member States. The savings of such economies of scale could then result in lower costs for industry and ultimately, the consumer. The savings achieved in some areas could be significant, given that the results of a recent study by PriceWaterhouseCoopers indicate that the cost of waste treatment varies substantially among the Member States (Price, 1998).

Utilisation of commercial markings at the European level

Introduction

One of the most difficult issues of packaging waste at the European level has been the problem of how to handle disposed packaging which is manufactured for sale in both the Member State of origin and in other parts of the Community. It seems logical that the packaging waste be processed in the Member State of sale and not be returned to the Member State of origin. Yet, who will pay for the costs of sorting and processing the waste?

In anticipation of the German Ordinance, *DSD* was created in late 1990 as an umbrella organisation to handle the collection and sorting of packaging, as well as to arrange for its treatment. This organisation began using *der Grüne Punkt* (the 'Green Dot') to identify packaging for which the manufacturer had paid a fee to *DSD.* Use of the Green Dot soon began to create problems: because products carrying the Green Dot were members of *DSD,* a retailer did not have to take back their packaging at the point of sale. Accordingly, retailers soon became reluctant to stock products which did not carry the Green Dot (Cairncross, 1993) and the situation continues to this day. Citing consumer preferences, German retailers refuse to allot shelf space to products whose packaging does not carry the Green Dot, which guarantees proper disposal. Manufacturers have effectively been forced to join the government-sponsored programme or else fail in the marketplace. The Industry Council for Packaging and the Environment was so concerned that exporters to Germany would have to join *DSD* in order to retain their market share, that it filed a complaint with the Commission (Mayer, 1991); the Commission then began an investigation, but did not take any action.

The Evolution of the Green Dot

Prior to the enactment of the Directive, France also began using a Green Dot for its *Eco-Emballages* system. Pursuant to an agreement between *Eco-Emballages* and *DSD,* manufacturers in France and Germany agreed to pay one fee to place the Green Dot on their packaging, thereby assuring compliance with both national systems. After the enactment of the Directive, *DSD* developed a plan to encourage the use of the Green Dot throughout the Community. This plan was enacted partly to facilitate the use of the Green Dot, which had become the standard in Germany, and partly in response to requests from manufacturers in other Member States, who preferred to proceed with an established symbol (Ringle 1996).

Accordingly, *DSD* established a new organisation, Packaging Recovery Organisation Europe (Pro-E). Based in Brussels, this organisation holds a license from *DSD* for the use of the Green Dot in Europe. The overall concept for licensing of the Green dot is as follows. Pro-E licenses the symbol and the trademark *'Der Grüne Punkt'* under main license contracts with the national packaging waste authorities in each Member States (for example, *Eco-Emballages* in France, *Duales System Deutschland* in Germany, and *Fost Plus* in Belgium). Thereafter, the national authority executes individual sub-licenses to each packaging manufacturer. The purpose of the Pro-E programme is to create a European-wide trademark which, when placed on a package, symbolises that the manufacturer of that product participates financially in a packaging waste collection and recovery system that complies with the Directive.

Interestingly, the trademark itself remains the property of *DSD* (and not Pro-E). *DSD* licenses use of the trademark to Pro-E and waives its right to directly license the trademark to the national packaging authorities. As *DSD* is the trademark owner, it bears all responsibility for the protection and defence of the trademark. However, Pro-E is required to reimburse *DSD* for all expenses incurred during a defence of the trademark (Ringle 1996). Clearly, should such an event occur, the costs of said defence would be passed down the trademark chain to the individual licensees and eventually, the consumer. Yet, none of the paying parties is allowed input in the trademark protection proceedings, such as whether to try the case in court or to settle it (which can make a substantial difference in the cost of a lawsuit).

The general license contract between *DSD* and Pro-E is for an unlimited duration and can only be terminated for 'important reasons'. The contract provides that upon termination, the parties will develop an arrangement for the continued use of the trademark for a 'reasonable period' (Ringle 1996). The exact licensing fee charged to Pro-E has not been made public. However, one can imagine the problems which arise when *DSD* raises the licensing fee. What can manufacturers do? They have already placed the Green Dot on their products and have helped to establish the trademark throughout Europe. The cost of changing to a new trademark will likely far exceed any proposed increase in price.

Since its inception, the Green Dot has encountered two problems, cross-border sales and free riders. A cross-border sale occurs when for example, a manufacturer in Aachen, Germany makes and sells a product that is then transported into Belgium, where the packaging is discarded. Ordinarily, the manufacturer must register and pay the fees of *DSD,* unless it can show that a substantial portion of its packaging is recovered in Belgium, in which case the manufacturer can pay the possibly lower fees to *Fost Plus. DSD* admits that it will discourage this type of activity, arguing that national laws must be respected. One should note however, that if for example, a German manufacturer makes packaging that will only be sold in France, it can register and pay the fees for a Green Dot from *Eco-Emballages.*

The second problem of free riders has created problems for Green Dot participants in two ways. First, some manufacturers either place a Green Dot on their product when they are not a member of the association or simply become members and then fail to pay their fees (Joly, 1996). A second problem occurs when manufacturers fail to participate in the programme, but the packaging waste from their products is nonetheless collected, sorted and treated. Although the first problem is easier to identify and correct, the second problem causes more difficulties (and greater expense) for the associations.

Criticisms of the Green Dot

With respect to Article 30 of the Treaty, it appears that the Green Dot acts a trade barrier, given that, for example, it is virtually impossible for a manufacturer to enter the German market without the Green Dot on its packaging, or that in certain Member States, such as France, the Green Dot is mandatory. However, complaints to this effect have not met with success at the Commission. Nonetheless, French retailers did have some success before their national courts in 1995, when Intermarché won a judgment against *Eco-Emballages.* The Court ruled that a retailer does not have to pay a fee for the Green Dot for its own house-brand products if the retailer's suppliers are already paying the fee (ERRA, 1998).

A second area of concern is the consumer mis-identification of the Green Dot as an 'eco-symbol', signifying a product which is environmentally friendly or which is made from recycled materials (Joly, 1996; de Sadeleer, 1995a). And a third concern is that the consumer will confuse the Green Dot with the markings used pursuant to the Directive to identify the packaging components and whether they are recyclable and/or reusable. In the end, it is possible that a package could contain symbols for not only the Green Dot and the Directive's composition markings, but also for the national symbol of ecological approval (such as Germany's Blue Angel) and, if marketed outside the Community, the international symbols for packaging composition! One begins to wonder if the time and expense of the markings is worth the potential environmental benefit thereof. The conflicts among the foregoing types of markings have contributed to the lack of progress in finalising the marking system required by the Directive. In the end however, the true problem of the 'different markings' is not whose marking will win out, but the apportionment of costs and controls among the authorities.

There are also competition issues caused by the use of the Green Dot. It is true that accords have been signed with the major packaging organisations in several Member States so that a firm which pays for the Green Dot in its own Member State will be able to sell its products in others. However, there will remain those small firms which cannot even afford the Green Dot in their home Member State. In effect, these firms will not be able to obtain shelf space at home or in other Member States without the Green Dot. The logistics and costs involved with the Green Dot thus discourage the smaller manufacturers from exporting, in direct contravention of one of the fundamental freedoms of the Treaty.

State Aids

Article 92 provides that any aid which is given by a Member State is an infringement of the Treaty if it distorts or threatens to distort competition. One first verifies that the state aid does not infringe other provisions of the Treaty; this being concluded, the state aid is next subject to a series of tests.

With regard to an economic instrument, the initial step is to determine if state aid is even involved. Although the proceeds from the economic instrument may constitute revenue for the Member State, it would not constitute a state aid unless some of the revenue was used to favour certain domestic enterprises or methods of production. The question would then arise as to how the economic instrument was designed and the extent to which national industries were involved in the deliberation process, as well as the intent of the legislator. The Commission has noted that exemptions from economic instruments can also be classified as state aid, even though the exemptions were granted to prevent the national firms from being disadvantaged, as compared with firms from countries where there are no such economic instruments (Commission, 1997).

Article 92(2) specifies types of aid which are compatible with the Treaty. As this paragraph does not specifically identify the protection of the environment, the only clearly compatible state aids regarding the environment would be those necessary as the result of severe environmental damage, i.e., a chemical spill, extensive forest fires or an 'act of nature'. On the other hand, Article 92(3) specifies aid which *may* be compatible; here, there is much more room for finding an environmental purpose. Article 92(3)(b) states that state aid may be granted to 'promote the execution of an important project of common European interest...'; this has been interpreted to give approval to concerted actions by Member States regarding aid for the environment. Moreover, Article 92(3)(c) provides that state aid may be awarded to 'facilitate the development of certain economic activities or of certain economic areas, where such aid does not adversely affect trading conditions to an extent contrary to the common interest'; this subparagraph has been used by the Commission to approve the majority of state aids for the environment (de Sadeleer, 1995a).

With respect to state aids for the environment, the guideline is that the benefits for the environment must outweigh the negative effects on competition. It is important to note that with the exception of minor aid which would fall under the *de minimis* rule, all state aid programmes must be notified to the Commission *prior* to implementation. If a Member State proceeds without notification, the aid is considered illegal; the Commission can order the Member State to recover the funds, with interest. Moreover, as the prohibition on distribution without notification has 'direct effect', a competitor of a firm receiving the illegal aid can apply to the national court for an injunction of payments or even an award of damages.

When analysing a prospective state aid, the Commission looks at both the source and the utilisation of the revenue. The ECJ has indicated that the Commission must consider all factors, including how the aid is financed; both indirect and direct aid; and the relationship between the financing and the sum that will be distributed as aid. The Commission has identified three ways in which funds from economic instruments are often employed as a form of state aid:

1. Support for investments or activities The Commission considers fact as such as whether the funds are spent in the same sector as they were collected, or whether another sector receives a benefit; whether the funds from the economic instrument will be sufficient for the activity or whether additional aid will be required; whether the funds given to firms are compensation for activities that they would not otherwise undertake and which activities are in the public interest; the intended duration of the economic instrument; and if the aid is to be reduced over the long-term.

2. Economic instruments for the collection and disposal of hazardous products/substances These are increasingly used; often the consumer pays an additional tax at the time of purchase and the funds are given to the firms to pay for the collection and disposal of the products after use. Programmes approved thus far have been for batteries, tyres and demolished automobiles.

3. Restibution to the group that paid the eco-tax With respect to products, these funds may not be redistributed to domestic producers to offset the burden of the tax. However, the funds may be used to address a domestic environmental problem caused by the product (Commission, 1997).

With regard to packaging waste, the issue of state aids would most likely arrise in the form of subsidies for a domestic recycling programme or eco-taxes which are used to benefit national manufacturers. As noted in Chapter 5, exceptions for domestic producers for economic instruments on products are very rare, as they tend to affect the free movement of goods and thus, must be very closely scrutinised. Although Member States are inclined to subsidise activities such as the collection, sorting and recycling of waste (de Sadeleer, 1995a), the Commission has not commenced any investigations in this area thus far. However, as the recycling and recovery industry further develops in the Community and market forces begin to have greater effects on the profitability of such operations, this area will bear closer scrutiny.

General competition policy issues

As the Commission noted in its recent communication on recycling, several problems are affecting the competitiveness of the European recycling market. The first of these is caused by the supply of recycled materials, which is facing competition from corresponding virgin raw materials, while being burdened with increased collection and sorting costs and the still inconsistent quality of recycled materials. A second problem for recyclers is the demand side, in that there is a lack of information in the customer base, which has led to misconceptions regarding the quality of recycled materials and to specifications for projects that actually discriminate against recycled materials. In certain Member States, the geographical remoteness has made it difficult to develop resale markets for recycled materials. The Commission is exploring solutions for redressing these issues, such as the enactment of standards for recycled materials and the creation of a Recycling Forum to encourage an exchange of information (EIS, 1998b; ERRA, 1998).

Conclusions

Overall, one can say that the initial stage of implementation of the Directive is finished: all Member States have now transposed some form of the Directive into their national legislation and have established some type of return and recovery system. The Directive is now moving into its second phase, during which standards will be established and Member States will provide reports on the quantities of waste collected and treated. By 30 June 2000, the Commission must to use this waste information to prepare a report which the Council will use to assess the entire packaging programme and to set targets for the next five-year period. Although the Council is supposed to set the new targets by 1 January 2001, given the contentiousness of the Directive's history, this date will likely be delayed. Nonetheless, the deadline will force the Community institutions, industry and NGOs to take a hard look at the work completed to date and to decide whether the environmental gains achieved thus far have been worth the cost, both financially and to the environment.

When one looks back at the last decade of packaging issues, from the initial discussions of an EC directive in the late 1980s to the enactment of the German Ordinance, from the enactment of the Directive itself to its subsequent implementation problems, one begins to ask whether the environment is any cleaner and whether the results are worth the cost. Aesthetically, the Directive has resulted in a cleaner environment, as it has helped to reduce one of the most visual aspects of society's waste. The Directive has forced companies to rethink how they package their products and has forced consumers to confront how much waste they produce on a regular basis. The Directive has also raised sovereignty issues regarding the right and extent to which a Member State may protect its environment. These legal issues will be a continuing source of debate for years to come, as the various parties argue whether the true purpose of the Directive is to assure the free flow of goods or to protect the environment.

Appendix A - The Packaging Waste Directive

EUROPEAN PARLIAMENT AND COUNCIL DIRECTIVE 94/62/EC[1]
of 20 December 1994 on packaging and packaging waste

THE EUROPEAN PARLIAMENT AND THE COUNCIL OF THE EUROPEAN UNION

Having regard to the Treaty establishing the European Community, and in particular Article 100a thereof,

Having regard to the proposal from the Commission,[2]

Having regard to the opinion of the Economic and Social Committee,[3]

Acting in accordance with the procedure laid down in Article 189b of the Treaty,[4]

Whereas the differing national measures concerning the management of packaging and packaging waste should be harmonized in order, on the one hand, to prevent any impact thereof on the environment or to reduce such impact, thus providing a high level of environmental protection, and, on the other hand, to ensure the functioning of the internal market and to avoid obstacles to trade and distortion and restriction of competition within the Community;

Whereas the best means of preventing the creation of packaging waste is to reduce the overall volume of packaging;

Whereas it is important, in relation of the objectives of this Directive, to respect the general principle that measures taken in one Member State to protect the environment should not adversely affect the ability of other Member States to achieve the objectives of the Directive;

Whereas the reduction of waste is essential for the sustainable growth specifically called for by the Treaty on European Union;

Whereas this Directive should cover all types of packaging placed on the market and all packaging waste; whereas therefore, Council Directive

85/339/EEC of 27 June 1985 on containers of liquids for human consumption[5] should be repealed;

Whereas packaging has a vital social and economic function and therefore measures provided for in this Directive should apply without prejudice to other relevant legislative requirements affecting the quality and transport of packaging or packaged goods;

Whereas, in line with the Community strategy for waste management set out in Council resolution of 7 May 1990 on waste policy[6] and Council Directive 75/442/EEC of 15 July 1975 on waste,[7] the management of packaging and packaging waste should include as a first priority, prevention of packaging waste and, as additional fundamental principles, reuse of packaging, recycling and other forms of recovering packaging waste and, hence, reduction of the final disposal of such waste;

Whereas, until scientific and technological progress is made with regard to recovery processes, reuse and recycling should be considered preferable in terms of environmental impact; whereas this requires the setting up in the Member States of systems guaranteeing the return of used packaging and/or packaging waste; whereas life-cycle assessments should be completed as soon as possible to justify a clear hierarchy between reusable, recyclable and recoverable packaging;

Whereas prevention of packaging waste shall be carried out through appropriate measures, including initiatives taken within the Member States in accordance with the objectives of this Directive;

Whereas Member States may encourage, in accordance with the Treaty, reuse systems of packaging which can be reused in an environmentally sound manner, in order to take advantage of the contribution of such systems to environmental protection;

Whereas from an environmental point of view, recycling should be regarded as an important part of recovery with a particular view to reducing the consumption of energy and of primary raw materials and the final disposal of waste;

Whereas energy recovery is one effective means of packaging waste recovery;

Whereas targets set in Member States for the recovery and recycling of packaging waste should be confined within certain ranges so as to take account of the different situations in Member States and to avoid creating barriers to trade and distortion of competition;

Whereas, in order to achieve results in the medium term and to give economic operators, consumers and public authorities the necessary perspective for the longer term, a medium-term deadline should be set for attaining the aforementioned targets and a long-term deadline set for targets to be determined at a later stage with a view to substantially increasing those targets;

Whereas the European Parliament and the Council should, on the basis of reports by the Commission, examine the practical experience gained in

Member States in working towards the aforementioned targets and the findings of scientific research and evaluation techniques such as eco-balances;

Whereas Member States which have, or will set, programmes going beyond such target ranges should be permitted to pursue those targets in the interest of a high level of environmental protection on condition that such measures avoid disturbances on the internal market and do not prevent other Member States from complying with this Directive; whereas the Commission should confirm such measures after appropriate verification;

Whereas, on the other hand, certain Member States may be allowed to adopt lower targets because of the specific circumstances in those Member States, on condition that they achieve a minimum target for recovery within the standard deadline, and the standard targets by a later deadline;

Whereas the management of packaging and packaging waste requires the Member States to set up return, collection and recovery systems; whereas such systems should be open to the participation of all interested parties and be designed to avoid discrimination against imported products and barriers to trade or distortions of competition and to guarantee the maximum possible return of packaging and packaging waste, in accordance with the Treaty;

Whereas the issue of Community marking of packaging requires further study, but should be decided by the Community in the near future;

Whereas, in order to minimise the impact of packaging and packaging waste on the environment and to avoid barriers to trade and distortion of competition, it is also necessary to define the essential requirements governing the composition and the reusable and recoverable (including recyclable) nature of packaging;

Whereas the presence of noxious metals and other substances in packaging should be limited in view of their environmental impact (in particular in the light of their likely presence in emissions or ash when packaging is incinerated, or in leachate when packaging is landfilled); whereas it is essential, as a first step towards reducing the toxicity of packaging waste, to prevent the addition of noxious heavy metals to packaging and ensure that such substances are not released into the environment, with appropriate exemptions which should be determined by the Commission in specific cases under a Committee procedure;

Whereas, if a high level of recycling is to be attained and health and safety problems are to be avoided by those employed to collect and process packaging waste, it is essential for such waste to be sorted at source;

Whereas the requirements for the manufacturing of packaging should not apply to packaging used for a given product before the date of entry into force of this Directive; whereas a transition period for the marketing of packaging is also required;

Whereas the timing of the provision on the placing on the market of packaging which meets all essential requirements should take account of the fact that European standards are being prepared by the competent standardization body; whereas, however, the provisions on means of proof of conformity of national standards should apply without delay;

Whereas the preparation of European standards for essential requirements and other related issues should be promoted;

Whereas the measures provided for in this Directive imply the development of recovery and recycling capacities and market outlets for recycled packaging materials;

Whereas the inclusion of recycled material in packaging should not contradict relevant provisions on hygiene, health and consumer safety;

Whereas Community-wide data on packaging and packaging waste are needed in order to monitor the implementation of the objectives of this Directive;

Whereas it is essential that all those involved in the production, use, import and distribution of packaging and packaged products become more aware of the extent to which packaging becomes waste, and that in accordance with the polluter-pays principle they accept responsibility for such waste; whereas the development and implementation of the measures provided for in this Directive should involve and require the close co-operation of all the partners, where appropriate, within a spirit of shared responsibility;

Whereas consumers play a key role in the management of packaging and packaging waste and thus have to be adequately informed in order to adapt their behaviour and attitudes;

Whereas the inclusion of a specific chapter on the management of packaging and packaging waste in the waste management plans required pursuant to Directive 75/442/EEC will contribute to the effective implementation of this Directive;

Whereas, in order to facilitate the achievement of the objectives of this Directive, it may be appropriate for the Community and the Member States to use economic instruments in accordance with the provisions of the Treaty, so as to avoid new forms of protectionism;

Whereas Member States should, without prejudice to Council Directive 83/189/EEC of 28 March 1983 laying down a procedure for the provision of information in the field of technical standards and regulations,[8] notify the Commission of drafts of any measures they intend to adopt before adopting them, so that it can be established whether or not they comply with the Directive;

Whereas the adaptation to scientific and technical progress of the packaging identification system and the formats relating to a database system should be ensured by the Commission under a committee procedure;

Whereas it is necessary to provide for specific measures to be taken to deal with any difficulties encountered in the implementation of this Directive in accordance, where appropriate, with the same committee procedure,

HAVE ADOPTED THIS DIRECTIVE:

Article 1

Objectives

1. This Directive aims to harmonize national measures concerning the management of packaging and packaging waste in order, on the one hand, to prevent any impact thereof on the environment of all Member States as well as of third countries or to reduce such impact, thus providing a high level of environmental protection, and, on the other hand, to ensure the functioning of the internal market and to avoid obstacles to trade and distortion and restriction of competition within the Community.
2. To this end this Directive lays down measures aimed, as a first priority, at preventing the production of packaging waste and, as additional fundamental principles, at reusing packaging, at recycling and other forms of recovering packaging waste and, hence, at reducing the final disposal of such waste.

Article 2

Scope

1. This Directive covers all packaging placed on the market in the Community and all packaging waste, whether it is used or released at industrial, commercial, office, shop, service, household or any other level, regardless of the material used.
2. This Directive shall apply without prejudice to existing quality requirements for packaging such as those regarding safety, the protection of health and the hygiene of the packed products or to existing transport requirements or to the provisions of Council Directive 91/689/EEC of 12 December 1991 on hazardous waste.[9]

Article 3

Definitions

For the purposes of this Directive:

1. 'packaging' shall mean all products made of any materials of any nature to be used for the containment, protection, handling, delivery and presentation of goods, from raw materials to processed goods, from the producer to the user or the consumer. 'Non-returnable' items used for the same purposes shall also be considered to constitute packaging.

   'packaging' consists only of:

   (a) sales packaging or primary packaging, i. e. packaging conceived so as to constitute a sales unit to the final user or consumer at the point of purchase;

(b) grouped packaging or secondary packaging, i.e., packaging conceived so as to constitute at the point of purchase a grouping of a certain number of sales units whether the latter is sold as such to the final user or consumer or whether it serves only as a means to replenish the shelves at the point of sale; it can be removed from the product without affecting its characteristics;

(c) transport packaging or tertiary packaging, i.e., packaging conceived so as to facilitate handling and transport of a number of sales units or grouped packagings in order to prevent physical handling and transport damage. Transport packaging does not include road, rail, ship and air containers;

2. 'packaging waste' shall mean any packaging or packaging material covered by the definition of waste in Directive 75/442/EEC, excluding production residues;

3. 'packaging waste management' shall mean the management of waste as defined in Directive 75/442/EEC;

4. 'prevention' shall mean the reduction of the quantity and of the harmfulness for the environment of:

- materials and substances contained in packaging and packaging waste,
- packaging and packaging waste at production process level and at the marketing, distribution, utilization and elimination stages, in particular by developing 'clean' products and technology;

5. 'reuse' shall mean any operation by which packaging, which has been conceived and designed to accomplish within its life cycle a minimum number of trips or rotations, is refilled or used for the same purpose for which it was conceived, with or without the support of auxiliary products present on the market enabling the packaging to be refilled; such reused packaging will become packaging waste when no longer subject to reuse;

6. 'recovery' shall mean any of the applicable operations provided for in Annex II.B to Directive 751442/EEC;

7. 'recycling' shall mean the reprocessing in a production process of the waste materials for the original purpose or for other purposes including organic recycling but excluding energy recovery;

8. 'energy recovery' shall mean the use of combustible packaging waste as a means to generate energy through direct incineration with or without other waste but with recovery of the heat;

9. 'organic recycling' shall mean the aerobic (composting) or anaerobic (biomethanization) treatment, under controlled conditions and using micro-organisms, of the biodegradable parts of packaging waste, which produces stabilized organic residues or methane. Landfill shall not be considered a form of organic recycling;

10. 'disposal' shall mean any of the applicable operations provided for in Annex II.A to Directive 75/442/EEC;

11. 'economic operators' in relation to packaging shall mean suppliers of packaging materials, packaging producers and converters, fillers and users, importers, traders and distributors, authorities and statutory organizations;

12. 'voluntary agreement' shall mean the formal agreement concluded between the competent public authorities of the Member State and the economic sectors concerned, which has to be open to all partners who wish to meet the conditions of the agreement with a view to working towards the objectives of this Directive.

Article 4

Prevention

1. Member States shall ensure that, in addition to the measures to prevent the formation of packaging waste taken in accordance with Article 9, other preventive measures are implemented. Such other measures may consist of national programmes or similar actions adopted, if appropriate, in consultation with economic operators, and designed to collect and take advantage of the many initiatives taken within Member States as regards prevention. They shall comply with the objectives of this Directive as defined in Article 1(1).

2. The Commission shall help to promote prevention by encouraging the development of suitable European standards, in accordance with Article 10.

Article 5

Member States may encourage reuse systems of packaging, which can be reused in an environmentally sound manner, in conformity with the Treaty.

Article 6

Recovery and recycling

1. In order to comply with the objectives of this Directive, Member States shall take the necessary measures to attain the following targets covering the whole of their territory;

   (a) no later than five years from the date by which this Directive must be implemented in national law, between 50 % as a minimum and 65 % as a maximum by weight of the packaging waste will be recovered;

   (b) within this general target, and with the same time limit, between 25 % as a minimum and 45 % as a maximum by weight of the totality of packaging materials contained in packaging waste will be recycled with a minimum of 15 % by weight for each packaging material:

(c) no later than 10 years from the date by which this Directive must be implemented in national law, a percentage of packaging waste will be recovered and recycled, which will have to be determined by the Council in accordance with paragraph 3(b) with a view to substantially increasing the targets mentioned in paragraphs (a) and (b).

2. Member States shall, where appropriate, encourage the use of materials obtained from recycled packaging waste for the manufacturing of packaging and other products.

3. (a) The European Parliament and the Council shall, on the basis of an interim report by the Commission, and four years from the date referred to in paragraph 1 (a) on the basis of a final report, examine the practical experience gained in the Member States in the pursuance of the targets and objective laid down in paragraphs 1 (a) and (b) and 2 and the findings of scientific research and evaluation techniques such as eco-balances.

(b) No later than six months before the end of the first five-year phase referred to in paragraph 1 (a) the Council shall, acting by qualified majority and on a proposal from the Commission, fix targets for the second five-year phase referred to in paragraph 1 (c). This process shall be repeated every five years thereafter.

4. The measures and targets referred to in paragraph 1 (a) and (b) shall be published by the Member States and shall be the subject of an information campaign for the general public and economic operators.

5. Greece, Ireland and Portugal may, because of their specific situation, i.e. respectively the large number of small islands, the presence of rural and mountain areas and the current low level of packaging consumption, decide to:

(a) attain, no later than five years from the date of implementation of this Directive, lower targets than those fixed in paragraph 1 (a) and (b), but shall at least attain 25 % for recovery;

(b) postpone at the same time the attainment of the targets in paragraph 1 (a) and (b) to a later deadline which, however, shall not exceed 31 December 2005.

6. Member States which have, or will, set programmes going beyond the targets of paragraph 1 (a) and (b) and which provide to this effect appropriate capacities for recycling and recovery, are permitted to pursue the targets in the interest of a high level of environmental protection, on condition that these measures avoid distortions of the internal market and do not hinder compliance by other Member States with the Directive. Member States shall inform the Commission thereof. The Commission shall confirm these measures, after having verified, in co-operation with the Member States, that they are consistent with the considerations above and do not constitute an arbitrary means of

discrimination or a disguised restriction on trade between Member States.

Article 7

Return, collection and recovery systems

1. Member States shall take the necessary measures to ensure that systems are set up to provide for:

   (a) the return and/or collection of used packaging and packaging waste from the consumer, other final user, or from the waste stream in order to channel it to the most appropriate waste management alternatives;

   (b) the reuse or recovery including recycling of packaging and/or packaging waste collected,

   in order to meet the objectives laid down in this Directive.

These systems shall be open to the participation of the economic operators of the sectors concerned and to the participation of the competent public authorities. They shall also apply to imported products under non-discriminatory conditions, including the detailed arrangements and any tariffs imposed for access to the systems, and shall be designed so as to avoid barriers l trade or distortions of competition in conformity with the Treaty.

2. The measures referred to in paragraph 1 shall form part of a policy covering all packaging and packaging waste and shall take into account, in particular, requirements regarding the protection of environment and consumer health, safety and hygiene; the protection of the quality, the authenticity and the technical characteristics of the packed goods and materials used and the protection of industrial and commercial property rights.

Article 8

Marking and identification system

1. The Council shall, in accordance with the conditions laid down in the Treaty, decide no latcr than two years after the entry into force of this Directive on the marking of packaging.

2. To facilitate collection, reuse and recovery including recycling, packaging shall indicate for purposes of its identification and classification by the industry concerned, the nature of the packaging material(s) used.

   To that end, the Commission shall, not later than 12 months after the entry into force of this Directive determine, on the basis of Annex I and in accordance with the procedure laid down in Article 21, the numbering and abbreviations on which the identification system is based and shall specify which materials shall be subject to the identification system in accordance with the same procedure.

3. Packaging shall bear the appropriate marking either on the packaging itself or on the label. It shall be clearly visible and easily legible. The marking shall be appropriately durable and lasting, including when the packaging is opened.

Article 9

Essential Requirements

1. Member States shall ensure that three years from the date of the entry into force of the Directive, packaging may be placed on the market only if it complies with all essential requirements defined by this Directive including Annex II.

2. Member States shall, from the date set out in Article 22(1), presume compliance with all essential requirements set out in this Directive, including Annex II in the case of packaging which complies:

   (a) with the relevant harmonised standards, the reference numbers of which have been published in the *Official Journal of the European Communities.* Member States shall publish the reference numbers of national standards transposing these harmonized standards;

   (b) with the relevant national standards referred to in paragraph 3 in so far as, in the areas covered by such standards, no harmonized standards exist.

3. Member States communicate to the Commission the text of their national standards, as referred to in paragraph 2(b), which they deem to comply with the requirements referred to in this Article. The Commission shall forward such texts forthwith to the other Member States.

   Member States shall publish the references of these standards. The Commission shall ensure that they are published in the *Official Journal of the European Communities.*

4. Where a Member State or the Commission considers that the standards referred to in paragraph 2 do not meet the essential requirements referred to in paragraph 1, the Commission or the Member State concerned shall bring the matter before the Committee set up by Directive 83/189/EEC giving the reasons therefor. This Committee shall deliver an opinion without delay.

   In the light of the Committee's opinion, the Commission shall inform Member States whether or not it is necessary to withdraw those standards from the publications referred to in paragraphs 2 and 3.

Article 10

Standardization

The Commission shall promote, as appropriate, the preparation of European standards relating to the essential requirements referred to in Annex II.

The Commission shall promote, in particular, the preparation of European standards relating to:

- criteria and methodologies for life-cycle analysis of packaging,
- the methods for measuring and verifying the presence of heavy metals and other dangerous substances in the packaging and their release into the environment from packaging and packaging waste,
- criteria for a minimum content of recycled material in packaging for appropriate types of packaging,
- criteria for recycling methods,
- criteria for composing methods and produced compost,
- criteria for the marking of packaging.

Article 11

Concentration levels of heavy metals present in packaging

1. Member States shall ensure that the sum of concentration levels of lead, cadmium, mercury and hexavalent chromium present in packaging or packaging components shall not exceed the following:
   - 600 ppm by weight two years after the date referred to in Article 22(i);
   - 250 ppm by weight three years after the date referred to in Article 22(i);
   - 100 ppm by weight five years after the date referred to in Article 22(i).
2. The concentration levels referred to in paragraph 1 shall not apply to packaging entirely made of lead crystal glass as defined in Directive 69/493/EEC.[10]
3. The Commission shall, in accordance with the procedure laid down in Article 21, determine:
   - the conditions under which the above concentration levels will not apply to recycled materials and to product loops which are in a closed and controlled chain,
   - the types of packaging which are exempted from the requirement referred to in paragraph 1, third indent.

Article 12

Information systems

1. Member States shall take the necessary measures to ensure that databases on packaging and packaging waste are established, where not already in place, on a harmonized basis in order to contribute to enabling Member States and the Commission to monitor the implementation of the objectives set out in this Directive.

2. To this effect, the databases shall provide in particular information on the magnitude, characteristics and evolution of the packaging and packaging waste flows (including information on the toxicity or danger of packaging materials and components used for their manufacture) at the level of individual Member States.

3. In order to harmonize the characteristics and presentation of the data produced and to make the data of the Member States compatible, Member States shall provide the Commission with their available data by means of formats which shall be adopted by the Commission one year from the date of entry into force of this Directive on the basis of Annex III, in accordance with the procedure laid down in Article 21.

4. Member States shall take into account the particular problems of small and medium-sized enterprises in providing detailed data.

5. The data obtained shall We made available with the national reports referred to in Article 17 and shall be updated in subsequent reports.

6. Member States shall require all economic operators involved to provide competent authorities with reliable data on their sector as required in this Article.

Article 13

Information for users of packaging

Member States shall take measures, within two years of the date referred to in Article 22 (1), to ensure that users of packaging, including in particular consumers, obtain the necessary information about:

- the return collection and recovery systems available to them,
- their role in contributing to reuse, recovery and recycling of packaging and packaging waste,
- the meaning of markings on packaging existing on the market,
- the appropriate elements of the management plans for packaging and packaging waste as referred to in Article 14.

Article 14

Management Plans

In pursuance of the objectives and measures referred to in this Directive, Member States shall include in the waste S management plans required pursuant to Article 17 of s Directive 75/442/EEC, a specific chapter on the management of packaging and packaging waste, including measures taken pursuant to Articles 4 and 5.

Article 15

Economics Instruments

Acting on the basis of the relevant provisions of the Treaty, the Council adopts economic instruments to promote the implementation of the objectives set by this Directive. In the absence of such measures, the Member States may, in accordance with the principles governing Community environmental policy, *inter alia,* the polluter-pays principle, and the obligations arising out of the Treaty, adopt measures to implement those objectives.

Article 16

Notification

1. Without prejudice to Directive 83/189/EEC, before adopting such measures, Member States shall notify the drafts of measures which they intend to adopt within the framework of this Directive to the Commission, excluding measures of a fiscal nature, but including technical specifications linked to fiscal measures which encourage compliance with such technical specifications, in order to permit the latter to examine them in the light of existing provisions following in each case the procedure under the above Directive.
2. If the proposed measure is also a technical matter within the meaning of Directive 83/189/EEC, the Member State concerned may indicate, when following the notification procedures referred to in this Directive, that the notification is equally valid for Directive 83/189/EEC.

Article 17

Obligation to report

Member States shall report to the Commission on the application of this Directive in accordance with Article 5 of Council Directive 91/692/EEC[11] standardizing and rationalizing reports on the implementation of certain Directives relating to the environment. The first report shall cover the period 1995 to 1997.

Article 18

Freedom to place on the market

Member States shall not impede the placing on the market of their territory of packaging which satisfies the provisions of this Directive.

Article 19

Adaptation to scientific and technical progress

The amendments necessary for adapting to scientific and technical progress the identification system - as referred to in Article 8 (2), Annex I and Article 10, last indent- and the formats relating to the database system - as referred to in Article 12 (3) and Annex III - shall be adopted in accordance with the procedure laid down in Article 21.

Article 20

Specific measures

1. The Commission, in accordance with the procedure laid down in Article 21, shall determine the technical measures necessary to deal with any difficulties encountered in applying the provisions of this Directive in particular to primary packaging for medical devices and pharmaceutical products, small packaging and luxury packaging.
2. The Commission shall also present a report to the European Parliament and the Council on any other measure to be taken, if appropriate accompanied by a proposal.

Article 21

Committee procedure

1. The Commission shall be assisted by a committee composed of the representatives of the Member States and chaired by the representative of the Commission.
2. The representative of the Commission shall submit to the committee a draft of the measures to be taken. The committee shall deliver its opinion on the draft within a time limit which the chairman may lay down according to the urgency of the matter. The opinion shall be delivered by the majority laid down in Article 148 (2) of the Treaty in the case of decisions which the Council is required to adopt on a proposal from the Commission. The votes of the representatives of the Member States within the committee shall be weighted in the manner set out in that Article. The chairman shall not vote.
3. (a) The Commission shall adopt the measures envisaged if they are in accordance with the opinion of the committee.

   (b) If the measures envisaged are not in accordance with the opinion of the committee, or if no opinion is delivered, the Commission shall, without delay, submit to the Council a proposal relating to the measures to be taken. The Council shall act by a qualified majority.

   If, on the expiry of a period which may in no case exceed three months from the date of referral to the Council, the Council has not acted, the proposed measures shall be adopted by the Commission.

Article 22

Implementation in national law

1. Member States shall bring into force the laws, regulations and administrative provisions necessary to comply with this Directive before 30 June 1996. They shall immediately inform the Commission thereof.
2. When Member States adopt these measures, they shall contain a reference to this Directive or shall be accompanied by such reference on the occasion of their official publication. The methods for making such reference shall be laid down by the Member States.
3. In addition, Member States shall communicate to the Commission all existing laws, regulations and administrative provisions adopted within the scope of this Directive.
4. The requirements for the manufacturing of packaging shall in no case apply to packaging used for a given product before the date of entry into force of this Directive.
5. Member States shall, for a period not exceeding five years from the date of the entry into force of the present Directive, allow the placing on the market of packaging manufactured before this date and which is in conformity with their existing national law.

Article 23

Directive 85/339/EEC is hereby repealed with effect from the date referred to in Article 22(1).

Article 24

This Directive shall enter into force on the day of its publication in the *Official Journal of the European Communities.*

Article 25

This Directive is addressed to the Member States.

Done at Brussels, 20 December 1994.

ANNEX I

IDENTIFICATION SYSTEM

The numbering used shall be from 1 to 19 for plastic, from 20 to 39 for paper and cardboard, from 40 to 49 for metal, from 50 to 59 for wood, from 60 to 69 for textiles, and from 70 to 79 for glass.

The identification system may also use the abbreviation for the relevant material(s) (e.g. HDPE: high density polyethylene). Materials may be identified by a numbering system and/or abbreviation. The identification marks shall appear in the centre of or below the graphical marking indicating the reusable or recoverable nature of the packaging.

ANNEX II

ESSENTIAL REQUIREMENTS ON THE COMPOSITION AND THE REUSABLE AND RECOVERABLE, INCLUDING RECYCLABLE, NATURE OF PACKAGING

1. Requirements specific to the manufacturing and composition of packaging

- Packaging shall be so manufactured that the packaging volume and weight be limited to the minimum adequate amount to maintain the necessary level of safety, hygiene and acceptance for the packed product and for the consumer.
- Packaging shall be designed, produced and commercialized in such a way as to permit its reuse or recovery, including recycling, and to minimize its impact on the environment when packaging waste or residues from packaging waste management operations are disposed of.
- Packaging shall be so manufactured that the presence of noxious and other hazardous substances and materials as constituents of the packaging material or of any of the packaging components is minimized with regard to their presence in emissions, ash or leachate when packaging or residues from management operations or packaging waste are incinerated or landfilled.

2. Requirements specific to the reusable nature of packaging

The following requirements must be simultaneously satisfied

- the physical properties and characteristics of the packaging shall enable a number of trips or rotations in normally predictable conditions of use;
- possibility of processing the used packaging in order to meet health and safety requirements for the workforce;
- fulfil the requirements specific to recoverable packaging when the packaging is no longer reused and thus becomes waste.

3. Requirements specific to the recoverable nature of packaging

(a) *Packaging recoverable in the form of recyclable material*

Packaging must be manufactured in such a way as to enable the recycling of a certain percentage by weight of the materials used into the manufacture of marketable products, in compliance with current standards in the Community. The establishment of this percentage may vary, depending on the type of material of which the packaging is composed.

(b) *Packaging recoverable in the form of energy recovery*

Packaging waste processed for the purpose of energy recovery shall have a minimum inferior calorific value to allow optimization of energy recovery.

(c) *Packaging recoverable in the form of composting*

Packaging waste processed for the purpose of composting shall be of such a biodegradable nature that it should not hinder the separate collection and the composting process or activity into which it is introduced.

(d) *Biodegradable packaging*

Biodegradable packaging waste shall be of such a nature that it is capable of undergoing physical, chemical, thermal or biological decomposition such that most of the finished compost ultimately decomposes into carbon dioxide, biomass and water.

ANNEX III

DATA TO BE INCLUDED BY MEMBER STATES IN THEIR DATABASES ON PACKAGING AND PACKAGING WASTE (IN ACCORDANCE WITH TABLES 1 TO 4)

1. For *primary, secondary and tertiary* packaging:
   (a) quantities, for each broad category of material, of packaging *consumed* within the country (produced + imported - exported) (Table 1);
   (b) quantities reused (Table 2).
2. For *household and non-household* packaging waste:
   (a) quantities for each broad category of material, *recovered and disposed of* within the country (produced + imported - exported) (Table 3);
   (b) quantities recycled and quantities *recovered* for each broad category of material (Table 4).

Table A.1[12]
Quantity of packaging (primary, secondary and tertiary) consumed within the national territory

	Tonnage produced	- Tonnage exported	+ Tonnage imported	= Total
Glass				
Plastic				
Paper/Cardboard (including composite)				
Metal				
Wood				
Other				
Total				

Table A.2
Quantity of packaging (primary, secondary and tertiary) reused within the national territory

	Tonnage of packaging consumed	Packaging reused	
		Tonnage	Percentage
Glass			
Plastic			
Paper/cardboard (including composite)			
Metal			
Wood			
Other			
Total			

Table A.3
Quantity of packaging waste recovered and disposed of within the national territory

	Tonnage of waste produced	- Tonnage of waste exported	+ Tonnage of waste imported	= Total
Household waste				
Glass packaging				
Plastic packaging				
Paper/cardboard packaging				
Cardboard composite packaging				
Metal packaging				
Wood packaging				
Total household packaging waste				
Non-household waste				
Glass packaging				
Plastic packaging				
Paper/cardboard packaging				
Cardboard composite packaging				
Metal packaging				
Wood packaging				
Total non-household packaging waste				

Table A.4
Quantity of packaging waste recycled or recovered within the national territory

	Total tonnage recovered and disposed of	Quantity recycled		Quantity recovered	
		Tonnage	Percentage	Tonnage	Percentage
Household waste					
Glass packaging					
Plastic packaging					
Paper/cardboard packaging					
Cardboard composite packaging					
Metal packaging					
Wood packaging					
Total household packaging waste					
Non-household waste					
Glass packaging					
Plastic packaging					
Paper/cardboard packaging					
Cardboard composite packaging					
Metal packaging					
Wood packaging					
Total non-household packaging waste					

Notes

1 *Official Journal,* 31 December 1994, L365.

2 *Official Journal* No. C 263, 12 October. 1992, p. 1 and *Official Journal* No. C 28, 21 October 1993, p. 1.

3 *Official Journal* No. C 129, 10 May 1993, p. 18.

4 Opinion of the European Parliament of 23 June 1993 *(Official Journal* No. C 194, 19 July 1993, p. 177), common position of the Council of 4 March 1994 *(Official Journal* No. C 137, 19 May 1994, p. 65) and Decision of the European Parliament of 4 May 1994 (*Official Journal* No. C 205, 25 July 1994, p. 163). Confirmed on 2 December 1993 (*Official Journal* No. C 342, 20 December 1993, p. 15). Joint text of the Conciliation Committee 8 November 1994.

5 *Official Journal* No. L 176, 6 July 1985, p. 18. Directive as amended by Directive 91/629/EEC *(Official Journal* No. L 377, 31 December 1991.

6 *Official Journal* No. C 122, 18 May 1990, p. 2.

7 *Official Journal* No. L 194, 25 July 1975, p. 39. Directive as amended by Directive 91/156/EEC *(Official Journal* No. L 78, 26 March 1991.

8 *Official Journal* No. L 109, 26 April 1983, p. 8. Directive as amended by Directive 92/400/EEC *(Official Journal* No. L 221, 6 August 1992.

9 *Official Journal* No. L 377, 31. December 1991, p. 20.

10 *Official Journal* No. L 326/36, 29 December 1969.

11 *Official Journal* No. L 377, 23 December 1991, p. 48.

12 (Emphasis added to original.) Table 1 is designed to determine the net tonnage of *packaging* consumed by considering the total *packaging* produced within the Member State, as well as the amount of *packaging* imported and exported. In Table 2, the net tonnage figure is then compared with the amount of *packaging* reused, in terms of total tonnage and as a percentage of the total *packaging* consumed.

Tables 3 and 4 are a sort of parallel to Tables 1 and 2 in that they determine the same figures, but for *packaging waste.* The categories are divided between Household and Non-household *packaging waste.* Table 3 is designed to determine the net total of *packaging waste* which is recovered and then disposed of within the Member State, i.e., *packaging waste* produced plus *packaging waste* imported, minus *packaging waste* exported. Table 4 then takes this net total and compares it with the quantities recycled and recovered, in terms of total tonnage and as a percentage of the total *packaging waste* produced.

Appendix B - Relevant Legislation

Extracts of Directive 75/442/EEC of 25 July 1975 on Waste (as Amended by Directive 91/156/EEC)

The categories defining waste are as follows:

- [C]onsumption residues not otherwise specified below
- Off-specification products
- Products whose date for appropriate use has expired
- Materials spilled, lost or having undergone other mishap, including any materials, equipment, etc., contaminated as a result of mishap
- Materials contaminated or soiled as a result of planned actions (e.g. residues from cleaning operations, packing materials, containers, etc.)
- Unusable parts (e.g. reject batteries, exhausted catalysts, etc.)
- Substances which no longer perform satisfactorily (e.g. contaminated acids, contaminated solvents, exhausted tempering salts, etc.)
- Residues of industrial processes (e.g. slags, still bottoms, etc.)
- Residues from pollution abatement processes (e.g. scrubber sludges, baghouse dusts, spent filters, etc.)
- Machining/finishing residues (e.g. lathe turnings, mill scales, etc.)
- Residues from raw materials extraction and processing (e.g. mining residues, oil field slops, etc.)
- Adulterated materials (e.g. oils contaminated with PCBs, etc.)
- Any materials, substances or products whose use has been banned by law
- Products for which the holder has no further use (e.g. agricultural, household, office, commercial and shop discards, etc.)
- Contaminated materials, substances or products resulting from remedial action with respect to land

- Any materials, substances or products which are not contained in the above categories.

Annex IIA of the Waste Directive lists the following operations:

- Tipping above or underground (e.g. landfill, etc.)
- Land treatment (e.g. biodegradation of liquid or sludge discards in soils, etc.)
- Deep injection (e.g. injection of pumpable discards into wells, salt domes or naturally occurring repositories, etc.)
- Surface impoundment (e.g. placement of liquid or sludge discards into pits, ponds or lagoons, etc.)
- Specially engineered landfill (e.g. placement into lined discrete cells which are capped and isolated from one another and the environment, etc.)
- Release of solid waste into a water body except seas/oceans
- Release into seas/oceans including seabed insertion
- Biological treatment... which results in final compounds or mixtures which are disposed of by means of any of the operations listed [herein]
- Physico-chemical treatment... which results in final compounds or mixtures which are disposed of by means of any of the operations [herein] (e.g. evaporation, drying, calcination, etc.)
- Incineration on land
- Incineration at sea
- Permanent storage (e.g. emplacement of containers in a mine, etc.)
- Blending of mixture prior to submission to any of the operations [herein]
- Repackaging prior to submission to any of the operations [herein]
- Storage pending any of the operations [herein] excluding temporary storage, pending collection, on the site where it is produced.

Annex II B of the waste Directive lists the following operations:

- Solvent reclamation/regeneration
- Recycling/reclamation of organic substances which are not used as solvents
- Recycling/reclamation of metals and metal compounds
- Recycling/reclamation of inorganic materials
- Regeneration of acids or bases
- Recovery of components used for pollution abatement
- Recovery of components from catalysts
- Oil re-refining or other reuses of oil
- Use principally as a fuel or other means to generate energy

- Spreading on land resulting in benefit to agriculture or ecological improvement, including composting and other biological transformation processes...
- Use of wastes obtained from an other the [above] operations
- Exchange of wastes for submission to any of the [above] operations
- Storage of materials intended for submission to any operation [herein], excluding temporary storage, pending collection, on the site where it is produced.

Appendix C - Relevant Commission Decisions

Commission Decision 97/129/EC establishing the identification system for packaging materials pursuant to European Parliament and Council Directive 94/62/EC on packaging and packaging waste
Official Journal n° L 50, 20/2/1997, p. 28.

THE COMMISSION OF THE EUROPEAN COMMUNITIES,

Having regard to the Treaty, establishing the European Community,
Having regard to European Parliament and Council Directive 94/62/EC of 20 December 1994 on packaging and packaging waste(1), and in particular Article 8(2) thereof,

Whereas the identification system is to be voluntary at least in a first stage but subject to revision to establish whether to introduce it on a binding basis at a further stage;
Whereas the identification system will be periodically reviewed and, if necessary, revised in accordance with the procedure laid down in Article 21 of Directive 94/62/EC;
Whereas the measures provided for in this Decision are in accordance with the opinion of the Committee established pursuant to Article 21 of Directive 94/62/EC;

HAS ADOPTED THIS DECISION:

Article 1
This Decision, which covers all packaging covered by Directive 94/62/EC, aims to establish the numbering and abbreviations on which the identification

system is based, indicating the nature of the packaging material(s) used, and specifying which materials shall be subject to the identification system.

Article 2

For the purposes of this Decision:

- the same definitions included in Article 3 of Directive 94/62/EC shall apply where relevant;
- composite: means packaging made up of different materials, and which cannot be separated by hand, none exceeding a given percent by weight which shall be established in accordance with the procedure laid down in Article 21 of Directive 94/62/EC. Potential exemptions for some materials will be established by the same procedure.

Article 3

The numbering and abbreviations of the identification system are as laid down in the Annexes.

Their use shall be voluntary for plastic materials mentioned in Annex I, the paper and fibreboard materials mentioned in Annex II, the metals mentioned in Annex III, the wood materials mentioned in Annex IV, the textile materials mentioned in Annex V, the glass materials mentioned in Annex VI, and the composites mentioned in Annex VII.

A decision whether to introduce on a binding basis the identification system for any material or materials may be adopted in accordance with the procedure laid down in Article 21 of Directive 94/62/EC.

Article 4

This Decision is addressed to the Member States.

Done at Brussels 28 January 1997
For the Commission
Ritt BJERREGAARD
Member of the Commission

(1) Official Journal n° L365, 31.12.1994, p. 10.

Annex I

Numbering and abbreviation system[1] for Plastics

Material	Abbreviation	Numbering
Polyethylene terephtalate	PET	1
High Density polyethylene	HDPE	2
Polyvinyl chloride	PVC	3
Low Density polyethylene	LDPE	4
Polypropylene	PP	5
Polystyrene	PS	6
[unassigned]		7-19

Annex II

Numbering and abbreviation system[1] for Paper and fibreboard

Material	Abbreviation	Numbering
Corrugated fibreboard	PAP	20
Non-corrugated fibreboard	PAP	21
Paper	PAP	22
[unassigned]		23-39

Annex III

Numbering and abbreviation system[1] for Metals

Material	Abbreviation	Numbering
Steel	FE	40
Aluminium	ALU	41
[unassigned]		42-49

Annex IV

Numbering and abbreviation system[1] for Wood materials

Material	Abbreviation	Numbering
Wood	FOR	50
Cork	FOR	51
[unassigned]		52-59

[1] Only capital letters shall be used.

Annex V

Numbering and abbreviation system[1] for Textile materials

Material	Abbreviation	Numbering
Cotton	TEX	60
Jute	TEX	61
[unassigned]		62-69

Annex VI

Numbering and abbreviation system[1] for Glass

Material	Abbreviation	Numbering
Colourless glass	GL	70
Green glass	GL	71
Brown glass	GL	72
[unassigned]		73-79

Annex VII

Numbering and abbreviation system[1] for Composite

Material	Abbreviation[2]	Numbering
Paper and fibreboard/miscellaneous metals		80
Paper and fibreboard/plastic		81
Paper and fibreboard/aluminium		82
Paper and fibreboard/tinplate		83
Paper and fibreboard/plastic/aluminium		84
Paper and fibreboard/plastic/aluminium/tinplate		85
[unassigned]		86-89
Plastic/aluminium		90
Plastic/tinplate		91
Plastic/miscellaneous metals		92
[unassigned]		93-94

---

[1] Only capital letters shall be used.

[2] Composites: cpt. Materials composites: c plus abbreviation corresponding to the predominant material (c/)

Glass/plastic	95
Glass/aluminium	96
Glass/tinplate	97
Glass/miscellaneous metals	98
[unassigned]	99

Commission Decision 97/622/EC concerning questionnaires for Member States reports on the implementation of certain Directives in the waste sector (Implementation of Council Directive 91/692/EEC)

Official Journal No. L 256, 19/09/1997, p. 13.

THE COMMISSION OF THE EUROPEAN COMMUNITIES,

Having regard to the Treaty establishing the European Community,
Having regard to Council Directive 91/692/EEC of 23 December 1991 on the standardization and rationalization of reports on the implementation of certain Directives relating to the Environment (1), and in particular Articles 5 and 6 and its Annex VI,
Having regard to Council Directive 75/442/EEC of 15 July 1975 on waste (2), as last amended by Commission Decision 96/350/EC (3),
Having regard to Council Directive 91/689/EEC of 12 December 1991 on hazardous waste (4), as last amended by Directive 94/31/EC (5),
Having regard to European Parliament and Council Directive 94/62/EC of 20 December 1994 on packaging and packaging waste (6),

Whereas Article 8(1) of Directive 91/689/EEC requires Member States to transmit to the Commission information on the application of this Directive in the context of the report provided for in Article 16(1) of Directive 75/442/EEC;
Whereas Article 17 of Directive 94/62/EC requires Member States to transmit to the Commission information on the application of this Directive in accordance with Article 5 of Directive 91/692/EEC;
Whereas Article 16 of Directive 75/442/EEC, has been replaced by Article 5 of Directive 91/692/EEC which requires Member States to transmit to the Commission information on the implementation of certain Community Directives in the form of a sectoral report;

Whereas, this report has to be established on the basis of a questionnaire or an outline drafted by the Commission in accordance with the procedure set out in Article 6 of Directive 91/692/EEC;
Whereas the first sectoral report will cover the period 1998 to 2000 inclusive;
Whereas the measures envisaged by this Decision are in accordance with the opinion expressed by the Committee established in accordance with Article 6 of the above mentioned Directive,

HAS ADOPTED THIS DECISION:

Article 1
The questionnaires attached to this Decision, which relate to Directive 91/689/EEC and Directive 94/62/EC, are hereby adopted.

Article 2
The Member States will use these questionnaires as a basis for the drawing up of the sectoral reports they are required to submit to the Commission pursuant to Article 5 of Directive 91/692/EEC and to Article 17 of Directive 94/62/EEC.

Article 3
This Decision is addressed to the Member States.

Done at Brussels, 27 May 1997.
For the Commission
Ritt BJERREGAARD
Member of the Commission

(1) OJ L 377, 31.12.1991, p. 48.
(2) OJ L 194, 25.7.1975, p. 39.
(3) OJ L 135, 6.6.1996, p. 32.
(4) OJ L 377, 31.12.1991, p. 20.
(5) OJ L 168, 2.7.1994, p. 28.
(6) OJ L 365, 31.12.1994, p. 10.

*   *   *

2. Questionnaire for the drafting of reports by Member States on the implementation and application of Directive 94/62/EC on packaging and packaging waste. There is no need to repeat information already supplied but please indicate clearly where and when the information was provided.

I. IMPLEMENTATION IN NATIONAL LAW

1. (a) Has the Commission been provided with details of the laws, regulations and administrative provisions introduced to comply with the Directive? (Yes/No)
   (b) If the answer to (a) is 'No', state the reason why.
2. (a) Are there plans to adopt further measures, not detailed below, within the framework of the Directive and under the scope of the notification obligation of Article 16? (Yes/No)
   (b) If the answer to (a) is 'Yes', has the Commission been notified of the measures in accordance with Article 16?
   (c) If the answer to (b) is 'No', state the reason why.
3. (a) If programs have been set up the objectives of which go beyond those referred to in Article 6 (1) (a) and (b), have those objectives been communicated to the Commission in accordance with Article 6 (6)? (Yes/No)
   (b) If the answer to (a) is 'No', state the reason why.

II. APPLICATION OF THE DIRECTIVE

1. (a) Have measures to prevent the formation of packaging waste been implemented in accordance with Article 4, other than those introduced in accordance with Article 9? (Yes/No)
   (b) If the answer to (a) is 'No', state the reason why.
   (c) Describe the measures taken, referring to any consultation with economic operators.
2. (a) Have measures been taken to encourage reuse systems in accordance with Article 5? (Yes/No)
   (b) If the answer to (a) is 'Yes', describe the measures.
3. (a) Have the necessary measures been taken, pursuant to Article 7, to set up systems for the return and/or collection of used packaging and/or packaging waste and systems for the reuse or recovery, including recycling, of packaging and/or collected packaging waste? (Yes/No)
   (b) If the answer to (a) is 'No', state the reason why.
   (c) Give details of the measures taken and of the systems which have been set up.
4. With regard to achievement of the recovery and recycling targets referred to in Article 6, please fill in and attach the formats adopted pursuant to Article 12 (3) indicating the method used to obtain data.

5. (a) Has the use of materials obtained from recycled packaging waste been encouraged in accordance with Article 6 (2)? (Yes/No)
   (b) If the answer to (a) is 'Yes', describe the actions undertaken.
6. How have the measures and targets referred to in Article 6 (1) (a) and (b) been published in accordance with Article 6(4)? Describe the information campaign aimed at the general public and economic operators.
7. What measures have been taken, in accordance with Article 13, to ensure that users of packaging receive the information laid down in that Article?
8. (a) Are there any national standards relating to the essential requirements, in accordance with Article 9, and to the concentration levels of heavy metals, in accordance with Article 11? (Yes/No)
   (b) If the answer to (a) is 'Yes', have they been communicated to the Commission? (Yes/No)
   (c) If the answer to (b) is 'No', state the reason why.
9. (a) Do the waste management plans required by Article 7 of Directive 75/442/EEC include a specific chapter on the management of packaging and packaging waste, in accordance with Article 14 of Directive 94/62/EC? (Yes/No)
   (b) If the answer to (a) is 'No', state the reason why.
10 (a) Have economic instruments been adopted, in accordance with Article 15, to attain the targets set in the Directive? (Yes/No)
   (b) If the answer to (a) is 'Yes', specify the measures adopted.

Commission Decision 97/138/EC establishing the formats relating to the database system pursuant to European Parliament and Council Directive 94/62/EC on packaging and packaging waste

Official Journal n° L 52, 22/02/1997, p. 22.

THE COMMISSION OF THE EUROPEAN COMMUNITIES,

Having regard to the Treaty establishing the European Community,
Having regard to European Parliament and Council Directive 94/62/EC of 20 December 1994 on packaging and packaging waste (1), and in particular Article 12(3) thereof,

Whereas the formats and the data framework should be periodically reviewed on the basis of practical experience, and if necessary, revised,
Whereas the measures provided for in this Decision are in accordance with the opinion of this Committee established pursuant to Article 21 of the Directive 94/62/EC,

HAS ADOPTED THIS DECISION:

Article 1
This Decision, which covers all packaging placed on the market in the Community and all packaging waste, as indicated in Article 2(1) of Directive 94/62/EC, aims to establish the formats relating to the database systems on packaging and packaging waste which are to be established in order to contribute to enabling Member States and the Commission to monitor the implementation of the objectives set out in Directive 94/62/EC. These formats are intended to harmonize the characteristics and presentation of the data produced and to make the data of the Member States compatible.

Article 2
For the purposes of this Decision:
- the same definitions included in Article 3 of Directive 94/62/EC shall apply where relevant.
- 'composite' means packaging made of different materials, and which cannot be separated by hand, none exceeding a given percent by weight, which shall be established in accordance with the procedure laid down in Article 21 of Directive 94/62/EC. Potential exemptions for some materials will be established by the same procedure.

Article 3
The formats presented in the Annexes shall be completed on an annual basis, starting with data for the year 1997 and covering the whole of each calendar year, and shall be provided to the Commission within 18 months of the end of the relevant year. They shall also be made available to the Commission with the national reports to be completed in accordance with Article 17 of Directive 94/62/EC.

Article 4
Member States shall present to the Commission appropriate qualitative information about concentration levels of heavy metals present in packaging in the sense of Article 11 of Directive 94/62/EC and on the presence of noxious and other hazardous substances and materials within the meaning of

the third indent of point 1 of Annex II to Directive 94/62/EC. Member States shall also present to the Commission quantitative information about packaging waste considered as hazardous due to contamination by product contents within the meaning of Council Directive 91/689/EEC (2) and Council Decision 94/904/EEC (3), in particular if it is not suitable for recovery. A report is to be presented to the Commission not later than the end of the first five year phase referred to in Article 6(1) of Directive 94/62/EC. This is to be repeated for the successive five-year periods.

Article 5

Member States shall present to the Commission the completed formats set out in this Decision, together with an appropriate description of how data have been compiled, as well as the main characteristics of the databases from which the data are extracted. In particular, the description shall include the estimations used in the calculation of the quantities and rates of packaging waste recovered and recycled and of the rates of reuse.

Article 6

The data to be included in Annex III (Tables 3, 4.1 and 4.2) concerning the weight of packaging waste recycled or recovered refer to the inputs of packaging waste to an effective recycling or recovery process. Only waste originating from packaging placed on the market may be considered for the calculation of these inputs, excluding any kind of production residue, from the production of packaging or of packaging materials or from any other process.

Article 7

The data contained in the formats are intended to monitor the implementation of the objectives of Directive 94/62/EC and serve also for information purposes and as a basis for future decision-taking. Annex II (Table 2) is to be completed on a voluntary basis.

The split of data in Annex III (Tables 3, 4.1, 4.2) on organic recycling, other forms of recycling, energy recovery and other forms of recovery, incineration and landfill shall be made solely for informative purposes and shall be on a voluntary basis. The provision of the data required in the columns headed 'total', 'total recycling' and 'total recovery' shall be obligatory. The provision of the data required in the column headed 'sorted for recycling' shall be voluntary. The packaging materials for which the provision of data is obligatory shall be glass, plastics, paper and fibreboard and metals.

Article 8

The Commission in accordance with the procedure laid down in Article 21 of Directive 94/62/EC, will review the framework for Member State provision of data, in order to make these data comparable and consistent. This framework should consider the definitions to be used, including composites, and the ranges of accuracy to be aimed for in the data. Member States shall ensure that the data provided comply with this framework.

Article 9

This decision is addressed to the Member States.

Done at Brussels 3 February 1997.
For the Commission,
Ritt BJERREGAARD,
Member of the Commission

Notes

(1) *Official Journal* No. L 365, 31.12.1994, p. 10.
(2) *Official Journal* No. L 377, 31.12.1991, p. 20.
(3) *Official Journal* No. L 356, 31.12.1994, p. 14.

Annex I

Table C.1
Quantity of packaging on the market within the Member State

Material		Production of empty packaging (tonnage)	Imports (empty packaging plus filled packaging) (tonnage)	Exports (empty packaging plus filled packaging) (tonnage)	Quantity placed on the market (tonnage)
Glass					
	PETP	✓✓✓✓✓✓✓	✓✓✓✓✓✓✓✓✓	✓✓✓✓✓✓✓✓✓	✓✓✓✓✓✓
	PE	✓✓✓✓✓✓✓	✓✓✓✓✓✓✓✓✓	✓✓✓✓✓✓✓✓✓	✓✓✓✓✓✓
	PVC	✓✓✓✓✓✓✓	✓✓✓✓✓✓✓✓✓	✓✓✓✓✓✓✓✓✓	✓✓✓✓✓✓
Plastic	PP	✓✓✓✓✓✓✓	✓✓✓✓✓✓✓✓✓	✓✓✓✓✓✓✓✓✓	✓✓✓✓✓✓
	PS	✓✓✓✓✓✓✓	✓✓✓✓✓✓✓✓✓	✓✓✓✓✓✓✓✓✓	✓✓✓✓✓✓
	Other	✓✓✓✓✓✓✓	✓✓✓✓✓✓✓✓✓	✓✓✓✓✓✓✓✓✓	✓✓✓✓✓✓
	Total				
Paper and cardboard					
	Steel	✓✓✓✓✓✓✓	✓✓✓✓✓✓✓✓✓	✓✓✓✓✓✓✓✓✓	✓✓✓✓✓✓
Metals	Aluminium	✓✓✓✓✓✓✓	✓✓✓✓✓✓✓✓✓	✓✓✓✓✓✓✓✓✓	✓✓✓✓✓✓
	Total				
Composites		✓✓✓✓✓✓✓	✓✓✓✓✓✓✓✓✓	✓✓✓✓✓✓✓✓✓	✓✓✓✓✓✓
Wood		✓✓✓✓✓✓✓	✓✓✓✓✓✓✓✓✓	✓✓✓✓✓✓✓✓✓	✓✓✓✓✓✓
Others		✓✓✓✓✓✓✓	✓✓✓✓✓✓✓✓✓	✓✓✓✓✓✓✓✓✓	✓✓✓✓✓✓
TOTAL					

Annex II

Table C.2
Reusable packaging

MATERIAL	Type of packaging	Product	Quantity of product in reusable packaging	Quantity of product in reusable and in same type one-way packaging	Units of reusable in circulation	Mean annual trip figure	Life-time	Units of same type one-way packaging put into the market
GLASS	Bottles	drinks						
		other						
	Containers	✓✓✓✓✓✓	✓✓✓✓✓✓	✓✓✓✓✓✓✓✓✓✓				
PLASTIC	Drums-barrels > 20 l.-< 250 l.	food						
		non-food						
	Barrels > 250 l.	food						
		non-food						
	Big bags	✓✓✓✓✓✓	✓✓✓✓✓✓	✓✓✓✓✓✓✓✓✓✓				
	Bottles	drinks						
		other						
	Boxes	✓✓✓✓✓✓	✓✓✓✓✓✓	✓✓✓✓✓✓✓✓✓✓				
	Containers	✓✓✓✓✓✓	✓✓✓✓✓✓	✓✓✓✓✓✓✓✓✓✓				
	Crates	✓✓✓✓✓✓	✓✓✓✓✓✓	✓✓✓✓✓✓✓✓✓✓				
	Pallets	✓✓✓✓✓✓	✓✓✓✓✓✓	✓✓✓✓✓✓✓✓✓✓				
FIBRE-BOARD	Boxes	✓✓✓✓✓✓	✓✓✓✓✓✓	✓✓✓✓✓✓✓✓✓✓				
	Barrels	✓✓✓✓✓✓	✓✓✓✓✓✓	✓✓✓✓✓✓✓✓✓✓				
	Containers	✓✓✓✓✓✓	✓✓✓✓✓✓	✓✓✓✓✓✓✓✓✓✓				
	Pallets	✓✓✓✓✓✓	✓✓✓✓✓✓	✓✓✓✓✓✓✓✓✓✓				
METALS Aluminium	Containers drums < 50 l.	food						
		non-food						
	Containers drums > 50 l.-< 300 l.	food						
		non-food						
METALS Steel	Containers drums < 50 l.	food						
		non-food						
	Containers drums > 50 l.-< 300 l.	food						
		non-food						
WOOD	Boxes	✓✓✓✓✓✓	✓✓✓✓✓✓	✓✓✓✓✓✓✓✓✓✓				
	Crates	✓✓✓✓✓✓	✓✓✓✓✓✓	✓✓✓✓✓✓✓✓✓✓				
	Drums	✓✓✓✓✓✓	✓✓✓✓✓✓	✓✓✓✓✓✓✓✓✓✓				
	Pallets	✓✓✓✓✓✓	✓✓✓✓✓✓	✓✓✓✓✓✓✓✓✓✓				
	Pallets-boxes	✓✓✓✓✓✓	✓✓✓✓✓✓	✓✓✓✓✓✓✓✓✓✓				

Table 2 is to be completed on a voluntary basis and intended only for those product and/or packaging categories which are considered by the national authorities relevant in the context of Article 5 of Directive 94/62/EC.

Accordingly, the columns on packaging types and products intended to encompass the potential relevant ones in the reuse area, but only those items pertinent in the context of the national reuse systems are to be filled. If needed the headings may be adapted to the actual systems.

If the data is available, the general items drinks/food/non-food may be split into generic elements such as mineral water, soft drinks, milk, alcoholic beverages, meat, fish, detergent powder, etc.

The data to be provided and their accuracy should be in line with their availability and the costs involved and may be adapted to Member States' situations.

NOTES: Blacked out boxes are considered not relevant in all cases. Quantities relating to drinks/liquids will be given in litres and in kilograms in all other cases.

Annex III

Table C.3
Quantities of packaging waste (in tonnes) arising and managed within the Member State

	Total	Sorted for recycling	Recovered						Disposed by	
			Organic recycling	Other forms of recycling	Total recycling	Energy recovery	Other forms of recovery	Total recovery	Incineration	Landfill
GLASS		✓✓✓✓✓✓✓		✓✓✓✓✓			✓✓✓✓			✓✓✓✓✓✓
PET	✓✓✓	✓✓✓✓✓✓✓		✓✓✓✓✓	✓✓✓✓	✓✓✓✓	✓✓✓✓	✓✓✓✓	✓✓✓✓✓✓	✓✓✓✓✓✓
PE	✓✓✓	✓✓✓✓✓✓✓		✓✓✓✓✓	✓✓✓✓	✓✓✓✓	✓✓✓✓	✓✓✓✓	✓✓✓✓✓✓	✓✓✓✓✓✓
PVC	✓✓✓	✓✓✓✓✓✓✓		✓✓✓✓✓	✓✓✓✓	✓✓✓✓	✓✓✓✓	✓✓✓✓	✓✓✓✓✓✓	✓✓✓✓✓✓
PLASTIC	✓✓✓	✓✓✓✓✓✓✓		✓✓✓✓✓	✓✓✓✓	✓✓✓✓	✓✓✓✓	✓✓✓✓	✓✓✓✓✓✓	✓✓✓✓✓✓
PS	✓✓✓	✓✓✓✓✓✓✓		✓✓✓✓✓	✓✓✓✓	✓✓✓✓	✓✓✓✓	✓✓✓✓	✓✓✓✓✓✓	✓✓✓✓✓✓
Other	✓✓✓	✓✓✓✓✓✓✓		✓✓✓✓✓	✓✓✓✓	✓✓✓✓	✓✓✓✓	✓✓✓✓	✓✓✓✓✓✓	✓✓✓✓✓✓
Total		✓✓✓✓✓✓✓		✓✓✓✓✓		✓✓✓✓	✓✓✓✓		✓✓✓✓✓✓	✓✓✓✓✓✓
PAPER & FIBREBOARD		✓✓✓✓✓✓✓	✓✓✓✓✓	✓✓✓✓✓		✓✓✓✓	✓✓✓✓		✓✓✓✓✓✓	✓✓✓✓✓✓
Aluminium	✓✓✓	✓✓✓✓✓✓✓		✓✓✓✓✓	✓✓✓✓	✓✓✓✓	✓✓✓✓	✓✓✓✓		✓✓✓✓✓✓
METALS	✓✓✓	✓✓✓✓✓✓✓		✓✓✓✓✓	✓✓✓✓		✓✓✓✓	✓✓✓✓		✓✓✓✓✓✓
Steel	✓✓✓	✓✓✓✓✓✓✓		✓✓✓✓✓	✓✓✓✓		✓✓✓✓	✓✓✓✓		✓✓✓✓✓✓
Total		✓✓✓✓✓✓✓		✓✓✓✓✓			✓✓✓✓			✓✓✓✓✓✓
COMPOSITES	✓✓✓	✓✓✓✓✓✓✓	✓✓✓✓✓	✓✓✓✓✓	✓✓✓✓	✓✓✓✓	✓✓✓✓	✓✓✓✓	✓✓✓✓✓✓	✓✓✓✓✓✓
WOOD	✓✓✓	✓✓✓✓✓✓✓	✓✓✓✓✓	✓✓✓✓✓	✓✓✓✓	✓✓✓✓	✓✓✓✓	✓✓✓✓	✓✓✓✓✓✓	✓✓✓✓✓✓
OTHER	✓✓✓	✓✓✓✓✓✓✓	✓✓✓✓✓	✓✓✓✓✓	✓✓✓✓	✓✓✓✓	✓✓✓✓	✓✓✓✓	✓✓✓✓✓✓	✓✓✓✓✓✓
TOTAL		✓✓✓✓✓✓✓	✓✓✓✓✓	✓✓✓✓✓		✓✓✓✓	✓✓✓✓		✓✓✓✓✓✓	✓✓✓✓✓✓

Table C.4.1
Monitored quantities of packaging waste (in tonnes) arising within the Member State and recovered outside the Member State

	Recovered by					
	Organic recycling	Other forms of recycling	Total recycling	Energy recovery	Other forms of recovery	Total recovery
GLASS		✓✓✓✓✓✓✓✓✓			✓✓✓✓✓✓✓✓✓✓	
PET		✓✓✓✓✓✓✓✓✓	✓✓✓✓✓✓✓✓	✓✓✓✓✓✓✓✓✓✓	✓✓✓✓✓✓✓✓✓✓	✓✓✓✓✓✓
PE		✓✓✓✓✓✓✓✓✓	✓✓✓✓✓✓✓✓	✓✓✓✓✓✓✓✓✓✓	✓✓✓✓✓✓✓✓✓✓	✓✓✓✓✓✓
PVC		✓✓✓✓✓✓✓✓✓	✓✓✓✓✓✓✓✓	✓✓✓✓✓✓✓✓✓✓	✓✓✓✓✓✓✓✓✓✓	✓✓✓✓✓✓
PLASTIC PP		✓✓✓✓✓✓✓✓✓	✓✓✓✓✓✓✓✓	✓✓✓✓✓✓✓✓✓✓	✓✓✓✓✓✓✓✓✓✓	✓✓✓✓✓✓
PS		✓✓✓✓✓✓✓✓✓	✓✓✓✓✓✓✓✓	✓✓✓✓✓✓✓✓✓✓	✓✓✓✓✓✓✓✓✓✓	✓✓✓✓✓✓
Other		✓✓✓✓✓✓✓✓✓	✓✓✓✓✓✓✓✓	✓✓✓✓✓✓✓✓✓✓	✓✓✓✓✓✓✓✓✓✓	✓✓✓✓✓✓
Total		✓✓✓✓✓✓✓✓✓		✓✓✓✓✓✓✓✓✓✓	✓✓✓✓✓✓✓✓✓✓	
PAPER & FIBREBOARD	✓✓✓✓✓✓✓✓✓✓	✓✓✓✓✓✓✓✓✓		✓✓✓✓✓✓✓✓✓✓	✓✓✓✓✓✓✓✓✓✓	
Aluminium		✓✓✓✓✓✓✓✓✓	✓✓✓✓✓✓✓✓	✓✓✓✓✓✓✓✓✓✓	✓✓✓✓✓✓✓✓✓✓	✓✓✓✓✓✓
METALS		✓✓✓✓✓✓✓✓✓	✓✓✓✓✓✓✓✓		✓✓✓✓✓✓✓✓✓✓	✓✓✓✓✓✓
Steel		✓✓✓✓✓✓✓✓✓	✓✓✓✓✓✓✓✓		✓✓✓✓✓✓✓✓✓✓	✓✓✓✓✓✓
Total		✓✓✓✓✓✓✓✓✓			✓✓✓✓✓✓✓✓✓✓	
COMPOSITES	✓✓✓✓✓✓✓✓✓✓	✓✓✓✓✓✓✓✓✓	✓✓✓✓✓✓✓✓	✓✓✓✓✓✓✓✓✓✓	✓✓✓✓✓✓✓✓✓✓	✓✓✓✓✓✓
WOOD	✓✓✓✓✓✓✓✓✓✓	✓✓✓✓✓✓✓✓✓	✓✓✓✓✓✓✓✓	✓✓✓✓✓✓✓✓✓✓	✓✓✓✓✓✓✓✓✓✓	✓✓✓✓✓✓
OTHER	✓✓✓✓✓✓✓✓✓✓	✓✓✓✓✓✓✓✓✓	✓✓✓✓✓✓✓✓	✓✓✓✓✓✓✓✓✓✓	✓✓✓✓✓✓✓✓✓✓	✓✓✓✓✓✓
TOTAL	✓✓✓✓✓✓✓✓✓✓	✓✓✓✓✓✓✓✓✓		✓✓✓✓✓✓✓✓✓✓	✓✓✓✓✓✓✓✓✓✓	

Table C.4.2
Monitored quantities of packaging waste (in tonnes) arising outside the Member State and recovered outside the Member State

		Recovered by					
		Organic recycling	Other forms of recycling	Total recycling	Energy recovery	Other forms of recovery	Total recovery
GLASS		■	✓✓✓✓✓✓✓✓✓		■	✓✓✓✓✓✓✓✓✓	
PLASTIC	PET	■	✓✓✓✓✓✓✓✓✓	✓✓✓✓✓✓✓✓✓	✓✓✓✓✓✓✓✓✓	✓✓✓✓✓✓✓✓✓	✓✓✓✓✓✓✓
	PE	■	✓✓✓✓✓✓✓✓✓	✓✓✓✓✓✓✓✓✓	✓✓✓✓✓✓✓✓✓	✓✓✓✓✓✓✓✓✓	✓✓✓✓✓✓✓
	PVC	■	✓✓✓✓✓✓✓✓✓	✓✓✓✓✓✓✓✓✓	✓✓✓✓✓✓✓✓✓	✓✓✓✓✓✓✓✓✓	✓✓✓✓✓✓✓
	PP	■	✓✓✓✓✓✓✓✓✓	✓✓✓✓✓✓✓✓✓	✓✓✓✓✓✓✓✓✓	✓✓✓✓✓✓✓✓✓	✓✓✓✓✓✓✓
	PS	■	✓✓✓✓✓✓✓✓✓	✓✓✓✓✓✓✓✓✓	✓✓✓✓✓✓✓✓✓	✓✓✓✓✓✓✓✓✓	✓✓✓✓✓✓✓
	Other	■	✓✓✓✓✓✓✓✓✓	✓✓✓✓✓✓✓✓✓	✓✓✓✓✓✓✓✓✓	✓✓✓✓✓✓✓✓✓	✓✓✓✓✓✓✓
	Total	■	✓✓✓✓✓✓✓✓✓		✓✓✓✓✓✓✓✓✓	✓✓✓✓✓✓✓✓✓	
PAPER & FIBREBOARD		✓✓✓✓✓✓✓✓✓	✓✓✓✓✓✓✓✓✓		✓✓✓✓✓✓✓✓✓	✓✓✓✓✓✓✓✓✓	
METALS	Aluminium	■	✓✓✓✓✓✓✓✓✓	✓✓✓✓✓✓✓✓✓	✓✓✓✓✓✓✓✓✓	✓✓✓✓✓✓✓✓✓	✓✓✓✓✓✓✓
	Steel	■	✓✓✓✓✓✓✓✓✓	✓✓✓✓✓✓✓✓✓	■	✓✓✓✓✓✓✓✓✓	✓✓✓✓✓✓✓
	Total	■	✓✓✓✓✓✓✓✓✓		■	✓✓✓✓✓✓✓✓✓	
COMPOSITES		✓✓✓✓✓✓✓✓✓	✓✓✓✓✓✓✓✓✓	✓✓✓✓✓✓✓✓✓	✓✓✓✓✓✓✓✓✓	✓✓✓✓✓✓✓✓✓	✓✓✓✓✓✓✓
WOOD		✓✓✓✓✓✓✓✓✓	✓✓✓✓✓✓✓✓✓	✓✓✓✓✓✓✓✓✓	✓✓✓✓✓✓✓✓✓	✓✓✓✓✓✓✓✓✓	✓✓✓✓✓✓✓
OTHER		✓✓✓✓✓✓✓✓✓	✓✓✓✓✓✓✓✓✓	✓✓✓✓✓✓✓✓✓	✓✓✓✓✓✓✓✓✓	✓✓✓✓✓✓✓✓✓	✓✓✓✓✓✓✓
TOTAL		✓✓✓✓✓✓✓✓✓	✓✓✓✓✓✓✓✓✓		✓✓✓✓✓✓✓✓✓	✓✓✓✓✓✓✓✓✓	

NOTE on Tables C.3, C.4.1 and C.4.2

1. Data corresponding to Table C.3 may be split on a voluntary basis, into municipal and non-municipal.
2. The column 'total' is binding. The column 'sorted for recycling' is to be provided on a voluntary basis.
3. The columns 'organic recycling' and 'other forms of recycling' are to be provided on a voluntary basis. The column 'total recycling' is binding.
4. The columns 'energy recovery' and 'other forms of recovery' are to be provided on a voluntary basis. The column 'total recovery' is binding.
5. The columns titled 'incineration' and 'landfill' are to be provided on a voluntary basis.
6. The data referring to the split into different plastic categories, to the split of metals into steel and in aluminium, to the item on composites and to the item on wood, are to be provided on a voluntary basis.
7. Data on composites may be either included according to the predominant material by total weight or separately specified.
8. Black boxes are considered not relevant in all cases. Shaded [check-marked] boxes are to be completed on a voluntary basis.

Bibliography

Alliance for Beverage Cartons and the Environment (1997), 'Eco-Balances, Packaging Policy & the EU Case Against Germany'.

Cairncross, Frances (ed) (29 May 1993), 'Waste and the Environment', *The Economist,* 1-25.

Dallemagne, Damien, Anna Devlin and Jonathan Shopley, (4 January 1993), 'The Changing Packaging Landscape in Europe', *Spectrum,* vol. 38, 1-21.

Demey, Thierry, Jean-Pierre Hannequart and Karine Lambert (1996), *Packaging Europe,* IBGE: Brussels.

Eden, Sally (Winter 1996), 'The Politics of Packaging in the UK: Business', *Environmental Politics,* vol. 5, no. 4; 632-653.

Emballages (November 1992), 'Un Entretien Exclusif avec Michel Desruet, Directeur de l'Environnement de L'Oréal', 16-17 [Emballages 1992a].

Emballages (December 1992), 95-96 [Emballages 1992b].

Emballages (February 1994), 'Emballages Européennes', 42.

ERRA (1996), 'The Packaging and Packaging Waste Directive Implementation'.

ERRA (1997), 'Key concerns about German 72% minimum refill quotas on drinks packaging'.

ERRA (1998), 'Packaging Waste Legislation in EU Member States and Neighbouring Countries'.

European Commission (1996), 'Communication on Environmental Agreements', COM (96) 561 final (27 November 1996).

European Commission (1997), Minutes of Commission Hearing of 13 June 1997 on the subject of German refill quotas.

European Commission (1997), 'Communication on Environmental Taxes and Charges in the Single Market', COM (97) 9 Final (26 March 1997).

European Commission (1997), Press Release, 17 October 1997 [1997a].

European Commission (1997), Press Release, 19 December 1997 [1997b].

European Commission (1998), Press Release, February 1998 [1998a].
European Commission (1998), Press Release, 6 April 1998 [1998b].
European Commission (1998), Press Release, 30 June 1998 [1998c].
European Information Service, 5 November 1993.
European Information Service, 1 March 1994 [EIS,1994a].
European Information Service, 31 March 1994 [EIS,1994b].
European Information Service, 17 May 1994 [EIS,1994c].
European Information Service, 31 May 1994 [EIS, 1994d].
European Information Service, 14 June 1994 [EIS, 1994e].
European Information Service, 12 July 1994 [EIS, 1994f].
European Information Service, 27 September 1994 [EIS, 1994g].
European Information Service, 25 October 1994 [EIS, 1994h].
European Information Service, 20 December 1994 [EIS, 1994i].
European Information Service, 10 January 1995.
European Information Service, 23 January 1996 [EIS, 1996a].
European Information Service, 16 April 1996 [EIS, 1996b].
European Information Service, 11 June 1996 [EIS, 1996c].
European Information Service, 23 July 1996 [EIS, 1996d].
European Information Service, 14 January 1997 [1997a].
European Information Service, 22 April 1997 [1997b].
European Information Service, 13 January 1998 [1998a].
European Information Service, 8 September 1998 [1998b].
European Information Service, 20 October 1998 [1998c].
Europen (1996), 'The Use of Life-cycle Assessments (LCA) as a Policy Tool in the Field of Packaging Waste Management' [Europen, 1996].
Europen (1997), 'Reuse Quotas and Product Specific Targets for Packaging'.
Europen (1998), 'Infringement 91/4489 German Packaging Ordinance (Refill Quota)' [Europen, 1998a].
Europen (1998), 'Infringement German Packaging Ordinance - Drinks Refill Quotas' [Europen, 1998b].
Europen (1998), 'German Packaging Ordinance - Environmental Objective' [Europen, 1998c].
Europen (1998), 'German Packaging Ordinance - Deposit on single-use drinks containers' [Europen, 1998d].
Europen (1998), 'Memo supporting the non conformity with Article 30 of the German refill quota (Packaging Ordinance)' [Europen, 1998e].
Europen (1998): Letter from Managing Director Julian Carroll to the European Commission [Europen, 1998f].
Federal Republic of Germany (1997), 'Defense of Reuse Quota'.
Gotschall, Mary G. (1996), 'Making Big Money from Garbage', *Columbia Journal of World Business,* 100-107.

Haverland, Markus (1998) *National Autonomy, European Integration and the Politics of Packaging Waste.* AWSB: Utrecht.

Hempen, Susanne and Reigle-Newland, Amy (1993), 'Overview on regulations with respect to packaging waste in EC-Member States and selected other countries', Report for the Second European Congress on Packaging and the Environment, Brussels, Belgium.

Hooghe, Marc (1993), 'Too little, too slow: the impact of European environmental legislation on the EC member states', *Planologisch Nieuws,* vol. 13, no. 2, 169-178.

Koeman, Niels S.F. (1993), 'Bilateral Agreements between Government and Industry in Dutch Environmental Law', *European Environmental Law Review,* 174-184.

Jackson, James O. (18 October 1993), 'World Class Litterbugs', *Time,* p. 80.

Joly, Hervé (1996), 'Les déchets d'emballage en Allemagne, enjeu économique et politique', *Problèmes économiques,* n°2.457.

Long, Antoinette and Tamsin Bailey (July 1997), 'The Single Market and the Environment: the European Union's Dilemma: the Example of the Packaging Directive', *European Environmental Law Review,* 214-219.

Mayer, Catherine (November 1991), 'Waging War on Waste', *International Management,* 65-67.

OECD (1993), 'Applying economic instruments to packaging waste: practical issues for product charges and deposit-refund systems', OECD: Paris.

OECD (1996), 'Five Waste Streams to Reduce', Report of the workshop held by the OECD in Washington, D.C. on 29-31 March 1995, OECD: Paris.

Porter & Perchard (1996), 'Environmental and public health aspects of reusable and disposable food service packaging'.

Price, Charles and Annette Hauff (January 1994), 'Eco-taxes in Belgium', *European Environmental Law Review,* 2-5.

Price Waterhouse Coopers (September 1998), 'The Facts: European Cost/ Benefit Perspective. Management Systems for Packaging Waste'.

Reid, Donald A. (September 1997), 'The Packaging and Packaging Waste Directive', *European Environmental Law Review,* 239-242.

Ringle, Wolfgang (1996), 'Pro-E (Packaging Recovery Organisation Europe)', report for the third Annual Conference on Packaging Waste Law, Brussels, Belgium.

de Sadeleer, Nicolas (1995), *Le Droit Communautaire et Les Dechets.* Bruylant: Brussels [de Sadeleer 1995a].

de Sadeleer, Nicolas (1995), 'Les emballages, l'environnement et le marché intérieur: un singulière trilogie', *Revue du Marché Unique Européen,* vol. 2, 87-119 [de Sadeleer 1995b].

Sinclair, Robert (October 1993). 'Report on the Channels for Recycling and the Problem of Markets for Recycled Materials', Report for the Second European Congress on Packaging and the Environment, Brussels, Belgium.

Steiner, Josephine (1998), *Textbook on EEC Law,* Blackstone: London.

Thieffry, Patrick & van Doorn, Philip (1993), 'La Protection de l'Environnement à l'Épreuve du Principe de Proportionnalité: Le Cloisonnement du Marché Commun par les Réglementations Nationales sur la Collecte des Emballages', *Journal of the Union of International Lawyers,* (November 1993).

UNESDA (June 1997), 'UNESDA's Comments on the German Packaging Ordinance: Reuse Quota'.